Grassroots Engagement and Social Justice through Cooperative Extension

TRANSFORMATIONS IN HIGHER EDUCATION: THE SCHOLARSHIP OF ENGAGEMENT

Grassroots Engagement and Social Justice through Cooperative Extension

Edited by Nia Imani Fields and Timothy J. Shaffer

Michigan State University Press | East Lansing

Michigan State University Press
East Lansing, Michigan 48823-5245

Library of Congress Cataloging-in-Publication Data
Names: Fields, Nia Imani, editor. | Shaffer, Timothy J., 1982– editor.
Title: Grassroots engagement and social justice through cooperative extension / edited by Nia Imani Fields and Timothy J. Shaffer.
Description: East Lansing : Michigan State University Press, [2022] |
Series: Transformations in higher education : the scholarship of engagement |
Includes bibliographical references.
Identifiers: LCCN 2021052800 | ISBN 9781611864274 (paperback) | ISBN 9781609176969 (PDF) | ISBN 9781628954647 (ePub) | ISBN 9781628964585 (Kindle)
Subjects: LCSH: United States. Extension Service—History. | United States. Cooperative State Research, Education, and Extension Service—History. | National Institute of Food and Agriculture (U.S.)—History. | Agricultural extension work—Social aspects—United States. | Agriculture—Study and teaching—Social aspects—United States. | State universities and colleges—United States—History. | Social justice and education—United States. | Democracy and education—United States.
Classification: LCC S544 .G73 2022 | DDC 630.71—dc23/eng/20211220
LC record available at https://lccn.loc.gov/2021052800

Book design by Preston Thomas
Cover design by Shaun Allshouse, www.shaunallshouse.com
Cover art by Good Studio, Adobe Stock

Michigan State University Press is a member of the Green Press Initiative and is committed to developing and encouraging ecologically responsible publishing practices. For more information about the Green Press Initiative and the use of recycled paper in book publishing, please visit www.greenpressinitiative.org.

Visit Michigan State University Press at *www.msupress.org*

Transformations in Higher Education: The Scholarship of Engagement

The Transformations in Higher Education: The Scholarship of Engagement book series is designed to provide a forum where scholars can address the diverse issues provoked by community-campus partnerships that are directed toward creating innovative solutions to societal problems. Numerous social critics and key national commissions have drawn attention to the pervasive and burgeoning problems of individuals, families, communities, economies, health services, and education in American society. Such issues as child and youth development, economic competitiveness, environmental quality, and health and health care require creative research and the design, deployment, and evaluation of innovative public policies and intervention programs. Similar problems and initiatives have been articulated in many other countries, apart from the devastating consequences of poverty that burdens economic and social change. As a consequence, there has been increasing societal pressure on universities to partner with communities to design and deliver knowledge applications that address these issues, and to cocreate novel approaches to effect system changes that can lead to sustainable and evidence- based solutions. Knowledge generation and knowledge application are critical parts of the engagement process, but so too are knowledge dissemination and preservation. The Transformations in Higher Education: The Scholarship of Engagement series was designed to meet one aspect of the dissemination/preservation dyad.

This series is sponsored by the National Collaborative for the Study of University Engagement (NCSUE) and is published in partnership with the Michigan State University Press. An external board of editors supports the NCSUE editorial staff in order to ensure that all volumes in the series are peer reviewed throughout the publication process. Manuscripts embracing campus-community partnerships are invited from authors regardless of discipline, geographic place, or type of transformational change accomplished. Similarly, the series embraces all methodological approaches from rigorous randomized trials to narrative and ethnographic studies. Analyses may span the qualitative to quantitative continuum, with particular emphasis on mixed-model approaches. However, all manuscripts must attend to detailing critical aspects of partnership development, community involvement, and evidence of program changes or impacts. Monographs and books provide ample space for authors to address all facets of engaged scholarship thereby building a compendium of praxis that will facilitate replication and generalization, two of the cornerstones of evidence-based programs, practices, and policies. We invite you to submit your work for publication review and to fully participate in our effort to assist higher education to renew its covenant with society through engaged scholarship.

Burton Bargerstock
Laurie Van Egeren
Hiram E. Fitzgerald

Contents

To my husband and two sons—my journey for social justice is because of you and for you. Gratitude to my village who keeps me grounded, loved, and lifted. Lastly, this piece aims to serve as a microphone for those voices that have been silenced for far too long. With love and power.

—Nia Imani Fields

To all those, past and present, who have worked for a more democratic society. May we recognize the importance of education in a democracy.

—Timothy J. Shaffer

Introduction

Timothy J. Shaffer and Nia Imani Fields

Vigorous reciprocity. That's how Ruby Green Smith described Extension's role and relationship with communities in her 1949 book *The People's Colleges*, a history of Extension's role at Cornell University throughout New York. In the preface to her book, Smith offered this strikingly simple yet profound way to think about Extension and its role in society: "There is vigorous reciprocity in the Extension Service because it is with the people, as well as 'of the people, by the people, and for the people'" (Smith, 1949, p. ix). She emphasized that this reciprocity impacted the way that Extension should be thought about because it was not only the knowledge coming from the university to the people, but also its reverse: "it carries from the people to their State Colleges practical knowledge whose workability has been tested on farms, in industry, in homes, and in communities. In ideal extension work, science and art meet life and practices" (Smith, 1949, p. ix). For Smith, the profundity of "vigorous reciprocity" was something realized through practice—through interactions and relationships involving people from the university and community.

In the preface to the republished 2013 edition of Smith's book, land-grant university scholar Scott Peters noted how "Smith refers to land-grant institutions as 'people's colleges.' This designation didn't (and doesn't) reflect an inherent and fixed fact. It represents an aspirational ideal. . . . The distinctive essence of the people's college ideal isn't access. And it isn't the 'application' of science. It's participation. It's the work of public engagement" (Peters, 2013a, p. viii). Peters's emphasis on the notion of work creating what Extension means highlights an important aspect that we will attend to in this book. However, "the dominant view of Extension's main purpose, today and during most of its history, is to address problems in democracy through the provision and application of scientific and technical knowledge and expertise" (Peters, Alter & Shaffer, 2018a, p. 24). Importantly, many people have at least partially defined the work of Extension expressly in aspirational terms, connecting work on "problems in democracy with problems of democracy" that cultivate democratic practices (Peters, Alter & Shaffer, 2018a, p. 24). It is that spirit of aspirational possibilities rooted in the work of Extension professionals engaging diverse individuals and communities that defines this volume.

We begin this book with this statement about Extension's role as articulated by Ruby Green Smith because it speaks to the broader notion of what it means when Smith referred to those in Extension having a faith and set of values that "recognize the human relationships that contribute to what the ancient Greeks called 'the good life'" (Smith, 1949, p. 544). In Peters's view, this statement conveys a "philosophy of life—and politics—that animates, guides, and grounds Smith's view of extension's public mission. . . . it suggests a way of telling extension's story that is decidedly different from the way it's commonly told" (Peters, 2013a, p. xiii). In short, Smith's account of what it meant to engage in that "vigorous reciprocity" is a reminder that Extension has deep roots in the practice of engaging others not only as recipients of knowledge or programs, but as persons who had an essential role in helping cultivate and sustain democracy. But, significantly, the sense of Extension's work being "of, by, and for" the people has met the reality of disparities based on factors such as race—a theme woven throughout this book. A second example can help highlight this reality.

From the mid-1930s through the early 1940s, Extension helped realize a community-based civic discussion project that engaged millions of Americans about topics relevant to their everyday lives. These included issues such as taxes, agricultural policy, environmental concerns, and differences between urban and rural populations. A critical element of this book highlights the systemic and cultural barriers that inhibited the more aspirational ideals of Extension expressed by Smith to address the problems *in* democracy while also engaging meaningfully the problems *of* democracy with many of America's own citizens. One striking example comes from a glimpse into the role that W. E. B. Du Bois had as part of Schools of Philosophy, an interdisciplinary professional learning opportunity for Extension professionals about how they could integrate the role of democratic discussion into their work. Extension's partnership with the U.S. Department of Agriculture (USDA) brought prominent scholars into conversation with rural audiences, sometimes highlighting the gap between the rhetoric of democracy and its practice (Jewett, 2013; Shaffer, 2018).

By the 1940s, Du Bois was viewed as a significant voice from the Black community and the broader American intellectual landscape, especially because of the popularity of his books *The Souls of Black Folk* (1903) and *Black Reconstruction* (1935). It was because of his notoriety that Du Bois was selected as one of the few Black speakers invited to be part of the Schools. Notably, he was still only invited to the racially segregated School of Philosophy organized by the USDA and Extension at Prairie View Normal and Industrial College in 1941. This School of Philosophy for Negro Agricultural Workers was held on January 23–25, 1941. Its theme was "What is a Desirable National Agricultural Program?," and it included presentations from a number of leaders from historically Black institutions exploring topics such as "What Can Discussion Contribute to a Better Understanding of the Present Situation?" and "Can Science Save Civilization?" Du Bois's lectures were titled "General Social Background of the Present Situation" and "Individualism, Democracy and Social Control" (Program—School of Philosophy for Negro Agricultural Workers, 1941).

Echoing themes from his 1940 book *Dusk of Dawn*, Du Bois emphasized the outsized role of agriculture because it was one of the few fields where "the Negro is the producer" and was not reliant upon "interests dominated almost entirely by whites" (Du Bois, 1940/2007, p. 106). During the School, Du Bois spoke of the centrality of slavery to agriculture in the modern world and, based partly on his prior research travel to places such as China, Germany, and Japan, noted that he saw a revolution of "those laboring classes" around the world—except in the United States. As Du Bois put it:

> In the United States we are talking as though the economic organization was just the same as it was before the beginning of this war and it was going to keep on the same sort of way—we are going to have the same profit system, we are going to have the same division of income, we are going to have the same worship of capital. Now, in no other country in the world is there any such attempt to deceive themselves—the people of the country. Everybody knows that the organization and control of capital is passing through a crucial phase. It is never going back to the old phase. Those workers in the organization of life, particularly in manufactures, are going to have a voice and they are going to have a decisive voice. (Du Bois, 1941b, p. 7)

He continued his lecture highlighting the global challenges to those working in agriculture, to those who labor but are largely exploited through economic means. As he put it, "The question is the question of income. How much are the farmers going to get. . . . Now for the first time in this generation, under the New Deal, we have a discussion of that. It is a discussion which still makes certain people exceedingly warm under the collar" (Du Bois, 1941b, p. 8).

It was through this segregated School of Philosophy about the future of American democracy and the role of rural people within it that Du Bois saw the opportunity for Extension agents to engage in that discussion about the economy and American democracy. As he said after his focus on the economic story and one's role within that, there is an importance to think about economics within the broader framework of democratic work: "We are talking about democracy. We can't open our mouths today without saying something about democracy, but we never say anything about democracy in Africa; and we don't say much about democracy in the South. We're taught democracy among those people who have the power, who have had the power who propose to keep the power" (Du Bois, 1941b, p. 7). The importance here is that Du Bois was inviting his audience to think of themselves, their colleagues, and their neighbors as actors in the democratic story. He concluded his remarks by returning to the intersection of power, income, and voice:

> But, as a matter of fact, we are coming to a time when we know perfectly well that that isn't true, that if you are going to have democracy in the world, that democracy is going to be applied to the matter of income. Somebody decides how much income each one of you is going to get; now it is the business of the people who are going to take part in democracy to have a voice in that decision. (Du Bois, 1941b, p. 9)

Shortly after his experience of speaking to a segregated group of Extension professionals, Du Bois wrote about the experience in the *New York Amsterdam Star-News*, an African American newspaper. He noted the importance of aligning critical thinking about a host of issues with the ability to engage in group discussion for the sake of our democracy. The content of his lecture was not clear in the newspaper article, but a takeaway aligned with his final lines of having people take part in democracy in order to have a voice in the decision-making process was clear: "I put it to you frankly friends of Chicago, Washington and Boston, can you afford to miss this sort of thing and continue to talk sincerely about race relations in America?" (Du Bois, 1941a).

These two vignettes—from Ruby Green Smith and W. E. B. Du Bois—highlight elements of the Extension story that shed light on the importance of not flattening the narrative about an organization familiar to most reading this book. The relational approach of Smith's "vigorous reciprocity" and Du Bois's comment about expressing one's view on democracy to a gymnasium full of African American Extension educators because of racial segregation offer reminders of being open to and learning from unfamiliar narratives that are part of the fabric of Extension—then and now. These stories reveal otherwise obscured or erased elements of this complex organization that continues to wrestle with the demands and expectations of serving and engaging all people. These are two stories about possibilities embedded within challenges. They are about the difficult work of engaging people in community, with the sense of "community" being radically diverse depending on the time and place. Throughout this book, the authors contribute to the Extension story—and thus higher education and the United States more generally—in ways that embody attempts to cultivate possibility.

• • •

Higher education's role in American society is a bit unclear and complex. Institutions of higher learning continue to be inequitably accessible for marginalized audiences and, in many cases, can serve as a microcosm of larger societal injustices. For many, higher education is seen as an opportunity to improve one's economic status. It serves as a seemingly necessary step for social mobility and increased opportunity to pursue interests of one's choosing. This attitude is increasingly moving beyond undergraduate education, as graduate degrees become necessary for many professions and positions. Yet connecting higher education to a notion of democracy and social justice is difficult. Especially if "democracy" transcends categories that define it simply as a political structure rather than a way of life and if "social justice" encompasses more than grassroots activism. David Mathews calls democracy "a way of living to maintain a good life in concert with others" (Mathews, 1999, p. 13), and Harry Boyte says a "democratic way of life" is "created through public, political work of the people" (Boyte, 2004, p. 93). For the purposes of this book, we think about democracy as public relationships within a pluralist society. Herein, we define social justice as the act of distributing power, resources, opportunity, societal benefits, and protection in a way that is equitable for all members of society (Baltimore Racial Justice Action, 2016; Fields, 2019).

Part of higher education's role is to help shape and inform civic engagement through the development of students' learning opportunities in classrooms and communities (Jacoby, 2009) as well as through engagement work on the part of academic professionals (Dzur, 2019). Much of the literature about higher education's civic mission focuses on the former rather than exploring the complexity of the latter. Those within colleges and universities are often perceived as "experts" who solve problems. They fix things. Their role is to provide information and resources for communities and elected officials to make informed decisions. In many ways this is true. But what's absent from such a definition of academic professionals is a view that situates land-grant institutions and those within them as civic actors. It also presupposes that academic professionals are meant to deposit information into students and communities rather than engage in meaningful dialogue whereby education serves as a process of conscientization (Freire, 1970, 1974).

Institutions of higher learning have long been less accessible to marginalized communities and in many ways have perpetuated oppression within society. In fact, these institutions were established to maintain European standards of scholarship, religion, and politics. The Morrill Land-Grant Acts of 1862 and 1890 established land-grant colleges to specialize in agriculture, engineering, and military science. These institutions were intended to reach audiences who had historically been excluded from higher education. Yet it's rarely discussed that the land used to establish these land-grant institutions came through the transfer or violent seizure of Indigenous land (Lee & Ahtone, 2020). Reverend Dr. Martin Luther King Jr. (2018, p. 25) also mentions the inequitable story of the land-grant system in discussing his reasons for the Poor People's Campaign:

> At the very same time that the government refused to give the Negro any land, through an act of Congress our Government was giving away millions of acres of land in the west and the midwest, which meant it was willing to undergird its white peasants from Europe with an economic floor. But not only did they give the land, they built land grant colleges with government money to teach them how to farm; not only that, they provided county agents to further their expertise in farming; not only that, they provided low interest rates in order that they could mechanize their farms; not only that today, many of these people are receiving millions of dollars in federal subsidies not to farm and they are the very people telling the Black man that he ought to lift himself by his own bootstraps. And this is what we are faced with, and this is the reality. Now, when we come to Washington in this campaign, we are coming to get our check.

In this speech, King refers to county agents who were funded by Cooperative Extension. Extension was formalized in 1914 through the Smith-Lever Act as a partnership involving the federal, state, and county governments through the USDA and land-grant universities, to apply research and provide education. This act expanded the reach of research-based information from being only accessible to university students to a wider range of youth and adults throughout the country in order to address problems related to agriculture, home economics, mechanic arts, youth development, and other related fields. Yet there are more nuanced aspects to the historical and contemporary elements of Extension's work that attend to democratic aspirations and ideals (Bailey, 1916; Peters, 2002, 2013b, 2017). This history, for all that is known and obscure, influences Extension's role in communities today. Over the last century, Extension has adapted to changing times and landscapes and now engages rural, suburban, and urban communities in grassroots education—including youth development, agriculture and food systems, environmental education and stewardship, and financial wellness and prosperity. The breadth of Extension's impact is notable, especially as it continues to very easily be thought of in overly simplistic ways such as being both exclusively rural and white. While those are essential elements to its identity and work, Extension is an educational institution and approach that has always, thinking back to Smith's (1949) account, provided a "relational, dialogic" model in addition to the transference of expertise. Archibald (2019, p. 30) notes that "the exchange of knowledge in the pursuit of community development is not some new fad or some contemporary reimagining of what Extension is 'really supposed to be.'"

The contemporary challenges we face invite us to think about the ways in which the Extension system can and should play a role in addressing social and cultural injustices through the cocreation of knowledge, leveraging expertise but not being limited in the ways that we think about the relationship between democracy and expertise. This book focuses on cooperative attempts at reaching for cultural competence both organizationally and through programmatic efforts that address social justice concerns in a variety of settings and contexts.

While rooted in an understanding of the history and development of Cooperative Extension, the book focuses on contemporary efforts to address systemic inequities in communities. The book offers an alternative to the "expert" model that would have Extension educators not involved in difficult and sometimes contentious community work, but rather views Extension professionals as being deeply engaged within the communities they work with to improve life—individually and at the community level (Peters, Alter & Shaffer, 2018b). Furthermore, this book highlights Extension's role and responsibility in culturally relevant community education that is rooted in democracy and social justice.

An Overview of the Book's Two Parts

In part 1, "Philosophical and Pedagogical Foundations for Engaging Diverse Audiences," the book highlights philosophical and pedagogical foundations for serving and being in partnership with diverse populations. Such foundations include antidiscrimination policy and procedures, the social ecological model, program planning, culturally relevant programming, and evaluation. The examples are rooted in scholarly practice in and with communities. The book begins with a historical look at injustice in education followed by a discussion of Extension's role and responsibility in community education. Extension has been in existence for over one hundred years and is deeply ingrained within the land-grant system.

In "Early Chapters of Extension: Land Grants, Segregation, and the Development of Democratic Programming," Timothy J. Shaffer, Scott J. Peters, and Maurice D. Smith Jr. highlight the significant legislative acts that shaped the land-grant system, and particularly Extension. This chapter also describes the disparities that exist within the 1890 land-grant system. The authors focus on the origins of program planning within Extension's early years because those first decades of existence determined how Extension would engage diverse publics through program planning, demonstrating the different opportunities—and distinct impacts because of those decisions—for citizens and their communities. As we continue to wrestle with questions about the purpose and impact of Extension, the first decades of its existence remind us of the diverse elements that came to shape this organization.

In "Moving Beyond the Minimum Standard: Nondiscrimination Regulations, Policies, and Procedures," Norman E. Pruitt and Latoya M. Hicks discuss the operating constructs of the land-grant system and related legal compliance and nondiscrimination policies. The authors challenge us to think beyond compliance and toward system-level changes that support an inclusive workforce with equitable access to Extension programs.

Nia Imani Fields and Fe Moncloa then transition the reader to "Culture and Culturally Relevant Programming," where they explore elements of culture and culturally relevant pedagogies within Extension. This chapter includes a case study that demonstrates culturally relevant practices in engaging Spanish-speaking immigrant Latinx youths.

The section concludes with "Cultural Competence in Evaluation: One Size Does Not Fit All." Megan H. Owens, Michelle Krehbiel, and Teresa McCoy remind readers that evaluation must be rooted in cultural competence, social justice, deliberative democracy, participatory research, and empowerment research to amplify the voices of underrepresented communities. Extension's reach and future is limited if vulnerable populations remain disengaged in evaluation and excluded from accepted forms of knowledge.

In part 2, "Grassroots Engagement in Practice," authors highlight programs of practice that demonstrate effective methods of developing and implementing community-based education based on topics such as food systems and nutrition, environmental justice, financial equity, community capital and social justice, and engaging marginalized audiences through positive youth development.

This section begins by exploring grassroots programs with food systems. In "The USDA and Land-Grant Extension: A Legacy of Continuing Inequities," Robert Zabawa and Lindsey Lunsford discuss structural policies and practices that have influenced Extension's funding disparities in supporting African American farming. The authors also explore the role of Extension in ensuring a culturally and ethnically appropriate food system.

The book then transitions to "Reaching Marginalized Audiences through Positive Youth Development Programming: Challenges When Influencing Constituents Do Not Agree that 4-H Should Be Open to All." Jeff Howard and Amanda Wahle introduce readers to the origins of the 4-H youth development program and how the program is currently postured to engage marginalized audiences. This chapter includes a case study exploring organizational challenges in sustaining inclusive practices to engage LGBTQ youth in 4-H programming.

Matt Calvert, Keith Nathaniel, and Manami Brown then explore the community capitals framework in "Developing Social Capital as a Conduit to Social Justice." This chapter focuses on social capital as an outcome of 4-H positive youth development and highlights a case study that aims to advance social justice outcomes.

In "Addressing Diversity at Multiple Levels of the Social Ecological Model" Katherine E. Soule and Shannon Klisch define the social ecological framework and describe its application to increase equity through Extension programs.

Sacoby Wilson, Helen Cheng, Davin Holen, Erin Ling, Daphne Pee, and Andrew Lazur then explore environmental justice in "An Introduction to Environmental Justice and Extension Programs Engaging Vulnerable Communities." The authors explore promising practices for environmental justice and community engagement along with three case studies related to water, Hurricane Sandy, and climate change.

This section concludes with "Programming to Address Building Financial Equity." Katherine E. Soule and Derrick Robinson highlight the interconnection between individual financial stories and the larger social structures within society. This chapter highlights inequitable structures that perpetuate poverty and how Extension can be leveraged to build financial equity within communities.

The book's conclusion functions as a bookend to this introduction. In it, we offer a vision for the future of Extension as we continue to reach for the cultural competence necessary to address issues of systemic injustice within the communities we serve and of which we are a part. Thinking about ourselves as members of the communities in which we live and work allows us to bring expertise alongside a commitment to the democratic practices of inclusion and engagement.

REFERENCES

Archibald, T. (2019). Whose Extension Counts? A Plurality of Extensions and Their Implications for Credible Evidence Debates. *Journal of Human Sciences and Extension*, 7(2), 22–35.

Bailey, L. H. (1916). *Ground-Levels in Democracy*. Ithaca, NY.

Baltimore Racial Justice Action. (2016). Our Definitions. November 30. http://bmoreantiracist.org/resources/our-definitions/.

Boyte, H. C. (2004). *Everyday Politics: Reconnecting Citizens and Public Life*. Philadelphia: University of Pennsylvania Press.

Du Bois, W. E. B. (1903). *The Souls of Black Folk: Essays and Sketches*. Chicago: A. C. McClurg.

———. (1935). *Black Reconstruction: An Essay Toward a History of the Part Which Black Folk Played in the Attempt to Reconstruct Democracy in America, 1860–1880*. New York: Harcourt, Brace and Company.

———. (1941a). As the Crow Flies. *New York Amsterdam Star-News*, February 8.

———. (1941b). The General Social Background of the Present Situation, January 23. In University of Massachusetts Amherst, Vol. W. E. B. Du Bois Papers Series 2. Speeches: Special Collections and University Archives.

———. (2007). *Dusk of Dawn: An Essay Toward an Autobiography of a Race Concept*. New York: Oxford University Press. (Original work published 1940).

Dzur, A. W. (2019). Conclusion: Sources of Democratic Professionalism in the University. In N. V. Longo & T. J. Shaffer (Eds.), *Creating Space for Democracy: A Primer on Dialogue and Deliberation in Higher Education* (pp. 285–294). Sterling, VA: Stylus.

Fields, N. I. (2019). Increasing your Cultural Awareness and Equity in Extension Programs. https://wellconnectedcommunities.extension.org/wp-content/uploads/2019/05/Course-Instructions-Increasing-Cultural-Awareness-and-Equity-in-Extension-Programs.pdf.

Freire, P. (1970). *Pedagogy of the Oppressed* (M. B. Ramos, Trans. 30th Anniversary ed.). New York: Continuum.

———. (1974). *Education for Critical Consciousness*. New York: Continuum.

Jacoby, B. (2009). Civic Engagement in Today's Higher Education: An Overview. In B. Jacoby & Associates (Eds.), *Civic Engagement in Higher Education: Concepts and Practices* (pp. 5–30). San Francisco: Jossey-Bass.

Jewett, A. (2013). The Social Sciences, Philosophy, and the Cultural Turn in the 1930s USDA. *Journal of the History of the Behavioral Sciences*, 49(4), 396–427.

King, M. L., Jr. (2018). The Three Evils. *Atlantic Monthly*, MLK Special Edition, 24–25.

Lee, R., & Ahtone, T. (2020). Land-Grab Universities: Expropriated Indigenous Land Is the Foundation of the Land-Grant University System. *High Country News*, 52(4), 32–45. https://www.hcn.org/issues/52.4/indigenous-affairs-education-land-grab-universities.

Mathews, D. (1999). *Politics for People: Finding a Responsible Public Voice* (2nd ed.). Urbana: University of Illinois Press.

Peters, S. J. (2002). Rousing the People on the Land: The Roots of the Educational Organizing Tradition in Extension Work. *Journal of Extension*, 40(3), 3FEA1. Available at http://archives.joe.org/joe/2002june/a2001.php.

———. (2013a). Preface. In R. G. Smith, *The People's Colleges: A History of the New York State Extension Service in Cornell University and the State, 1876–1948* (pp. viii–xx). Ithaca, NY: Cornell University Press.

———. (2013b). Storying and Restorying the Land-Grant System. In R. L. Geiger & N. M. Sorber (Eds.), *The Land-Grant Colleges and the Reshaping of American Higher Education* (pp. 335–353). New Brunswick, NJ: Transaction Publishers.

———. (2017). Recovering a Forgotten Lineage of Democratic Engagement: Agricultural and Extension Programs in the United States. In C. Dolgon, T. D. Mitchell & T. K. Eatman (Eds.), *The Cambridge Handbook of Service Learning and Community Engagement* (pp. 71–80). New York: Cambridge University Press.

Peters, S. J., Alter, T. R., & Shaffer, T. J. (2018a). Introduction: Making Democracy Work as It Should. In S. J. Peters, T. R. Alter & T. J. Shaffer (Eds.), *Jumping into Civic Life: Stories of Public Work from Extension Professionals* (pp. 13–36). Dayton, OH: Kettering Foundation Press.

———. (Eds.). (2018b). *Jumping into Civic Life: Stories of Public Work from Extension Professionals.* Dayton, OH: Kettering Foundation Press.

Program—School of Philosophy for Negro Agricultural Workers. (1941). Record Group 83: Records of the Bureau of Agricultural Economics (Entry 19, Box No. 595). National Archives College Park.

Shaffer, T. J. (2018). Thinking Beyond Food and Fiber: Public Dialogue and Deliberation in the New Deal Department of Agriculture. In A. B. Hoflund, J. Jones & M. C. Pautz (Eds.), *The Intersection of Food and Public Health: Examining Current Challenges and Solutions in Policy and Politics* (pp. 307–326). New York: Routledge.

Smith, R. G. (1949). *The People's Colleges: A History of the New York State Extension Service in Cornell University and the State, 1876–1948*. Ithaca, NY: Cornell University Press.

PART 1

Philosophical and Pedagogical Foundations for Engaging Diverse Audiences

Early Chapters of Extension: Land Grants, Segregation, and the Development of Democratic Programming

Timothy J. Shaffer, Scott J. Peters, and Maurice D. Smith Jr.

When we think about the history of Cooperative Extension, we do well to recognize the complexity of the effort to *extend* the university to individuals and community beyond campus. Significantly, there have been different ways in which the idea of the land-grant university has been understood and experienced, particularly through the efforts of the Cooperative Extension Service. This chapter will introduce the significant legislative acts that shaped the land-grant system and, particularly, Extension. It introduces the First and Second Morrill Acts, the Hatch Act, and the Smith-Lever Act—the legislative foundation of Extension. For this volume, it also attends to the often-overlooked elements that shape the 1890 land-grant institutions and the intentional efforts to engage African Americans and others marginalized in the United States. Finally, the chapter looks at program planning in the first decades of Extension—from 1914 into the 1940s, with the first thirty-five years as a marker and moment for reflection about the idea and practices of Extension. While it may not be apparent, the shifts and adjustments in Extension at that time can serve as a reminder for us to consider, and reconsider, the ways in which we engage individuals and communities in our work today.

To interpret these foundational elements that have shaped Extension's role in and beyond land-grant colleges and universities, it is essential to think about any and all of these actions—legislative or otherwise—through the lens of politics. Levine (2014, p. 29) offers useful language to frame how we think about any of these actions that shaped Extension's purpose: "The question is *what* should we do, not what should be done?" This question reminds us that our actions are political in the sense that we are civic actors; our actions have consequences. As this chapter highlights, history doesn't just *happen*. It becomes reality through the actions of individuals, groups, and institutions.

The Morrill Act of 1862 and Its Impact on Democracy

A common way to speak about the development and role of land-grant colleges and universities is to highlight some of the significant events that have helped to shape the identity of these institutions. We will follow in this tradition and then we will critique it, explaining why this is important in order to understand the complexity of these institutions as they came to exist and transform from the middle of the nineteenth century into the first decades of the twentieth.

In the early nineteenth century, the U.S. economy and culture were agricultural in focus, with 85 percent of the population (of European descent) living in rural communities along the East Coast (Eddy, 1957, p. 1). It is fitting that this information is found at the beginning of the first general account of the development of the land-grant movement, because the Morrill Act recognized agriculture as central to the development of the United States as it moved westward.

During the first half of the nineteenth century, there were two types of colleges and universities: publicly controlled and privately controlled. European universities where American professors had trained and taught previously greatly influenced these institutions. They were designed to serve a stratified society with limited democratic aspirations. College education was primarily reserved for "the leisure classes, the government leaders, and members of the professions" (Brunner, 1962, p. 1). In many ways, this meant upper-class white men. This classic interpretation highlighted that higher education institutions in the United States functioned in a similar fashion and maintained a classical curriculum, with only slight adaptations to the needs of a "pioneer people" (Brunner, 1962, p. 1). But more recent scholarship (e.g., Dorn, 2017; Geiger, 2000b) complicates this simplistic narrative, highlighting the diversity of institutions rather than seeing them only as private institutions and then public institutions. As Geiger (2000a, p. 2) put it, the familiar "narrative histories . . . have provided the only general treatments of nineteenth-century higher education." As we think about the roots of land-grant colleges and universities, it is important to recognize that they were part of a cultural climate that was increasingly embracing formal education to inform and shape all manners of life and explored questions about meaning and purpose of education for broader impact. Land-grant institutions are best thought of as part of the innovations that came from "colleges for women, scientific schools, and practical courses in engineering and agriculture" (Geiger, 2000a, p. 9) during the middle of the nineteenth century.

In the United States, the modern public American research institution traces its roots to a handful of universities: the Universities of Georgia, North Carolina, Vermont, South Carolina, and Virginia. But the "real signal of public commitment" to university-based research came from the Morrill Act of 1862 (Rhoten & Powell, 2011, p. 317). Justin Morrill, a representative and then senator from Vermont, sponsored the Land-Grant College Act, which was signed into law by President Abraham Lincoln on July 2, 1862.[1] Morrill had proposed similar legislation previously, eventually having it passed by Congress but vetoed by President James Buchanan in 1859. Morrill, who had no formal education beyond secondary school, believed education could provide people access to a better way of life and make them better citizens. In a speech in 1888 about the Land-Grant Act, Morrill said that "the fundamental idea was to offer an opportunity in every state for a liberal and larger education to larger numbers, not merely to those destined to sedentary professions, but to those needing higher instruction for the world's business, for the industrial pursuits and professions of life" (Morrill, 1888, p. 11).

The establishment of a national system of universities that blended liberal and practical education challenged the transplanted European approach to higher education. The Morrill Act gave each state thirty thousand acres of federal land for each member of the congressional delegation from that state. Millions of acres were distributed, enabling the founding of colleges in thirty states before 1900 (Cohen & Kisker, 2010, p. 115). That land was then sold to create endowments to support and maintain "at least one college where the leading object shall be, without excluding other scientific and classical studies, and including military tactics, to teach such branches of learning as are related to agriculture and the mechanic arts . . . in order to promote the liberal and practical education of the industrial classes in the several pursuits and professions in life" (Eddy, 1957, p. 3).

But this familiar retelling of the origin story of these institutions leaves out an essential element making this federal act possible: the expropriated land from tribal nations scattered throughout twenty-four Western states. As recent research suggests, "In all, the act redistributed nearly 11 million acres—an area larger than Massachusetts and Connecticut combined. But with a footprint broken up into almost 80,000 parcels of land, scattered mostly across 24 Western states, its place in the violent history of North America's colonization has remained comfortably inaccessible. . . . Altogether, the grants, when adjusted for inflation, were worth about half a billion dollars" (Lee & Ahtone, 2020, p. 34). The continued benefit from this land acquisition is not only part of the beginning of these institutions, but as noted in *High Country News*, "The money has been on the books ever since, earning interest, while a dozen or more of those universities still generate revenue from unsold lands. Meanwhile, Indigenous people remain largely absent from student populations, staff, faculty and even curriculum" (Lee & Ahtone, 2020, p. 43). If we think of this as only a sad chapter from long ago, then we are obfuscating the current implications of this land grab that became grants. This is an important element of the story of these significant institutions. What is colloquially referred to as "land grants" bears an important mark (or a stain?) that should be acknowledged when we think about what made the transformation of nineteenth-century America, in large part because of these institutions, possible.

Morrill argued that American society, and especially its economy, needed to address the changes taking place with increasing industrialization and to meet the agricultural demands of a booming population. Because the federal government did not have the ability to provide funds at that time, federal land acquired through expropriation became a cornerstone of the establishment of public higher education. In this sense, the government functioned more as a real estate promoter than a funding source (Eddy, 1957, p. 36). The "land grab" (Lee & Ahtone, 2020) history of the land-grant story and the ways in which tribal lands became intertwined with the sense of these public institutions that are commonly referred to as "land-grants" without a sense of the *land* highlights the importance of understanding history to inform the present.

The model of what land-grant colleges and universities could be was not without precedent. While states were charged with the task of creating new academic institutions, some had previously established colleges of agriculture. The precursor to Michigan State University often served as the example of what the land-grant colleges and universities could draw on to form a foundation of practical education (Clute, 1891). These newly created public education institutions were designed to be "elite without being elitist, to provide access to knowledge and education to those previously denied such access" (Simon, 2010, p. 100). Before this period, higher education was typically limited to white men from affluent backgrounds. These new institutions challenged social norms and expanded educational opportunities for diverse citizens.

The desire to educate people with scientific agricultural practices as well as afford the opportunity for practical learning to take place alongside liberal education speaks to a *political* question. This sense of land-grants being situated within a broader cultural and political context is critical if we think about the importance of reflecting on the issue of what citizens do as actors with agency in the world (Levine, 2014). As demonstrated in the 1890 Morrill Act, the sense of who was deemed acceptable for such educational opportunities was rooted in identity and race. The issue of race and acceptability played out in other states as well, not only in those that fought to retain slavery in the Southern states. An example from one land-grant institution will demonstrate this point.

Cornell University, New York's land-grant institution, was open to both men and women of all races and challenged many of the social conventions in the 1860s. Before World War I, the Cornell student body included "representatives from every quarter of the globe," international students from Canada, Mexico, Cuba, China, and Japan, among many others. When graduate students were taken into account, "a still greater diversity and considerably larger total would be manifest" (Von Engeln, 1924, p. 226). Told this way, Cornell appears to be welcoming of all peoples.

But in practice, women didn't come to Cornell's campus until 1870, and it was not until 1873 that the first woman graduated. In 1929, Ezra Cornell's founding belief of "any person any study" translated into Ruth Peyton, an African American undergraduate student, being denied residency in the women's dormitory because, as President Livingston Farrand wrote to Peyton's mother, "the placing of a colored student in one of the dormitories inevitably caused more embarrassment than satisfaction for such a student . . . while I have great sympathy for your feeling, I cannot order a change in the procedure of the Dean of Women, under whose jurisdiction the matter falls" (Farrand, 1929). There may have been students from around the world, but a student from Olean, New York, was discriminated against because of her skin color. The idea of "any person any study" was not quite the welcoming slogan it appeared to be. In short, higher education offered an opportunity for women and diverse populations, but institutions struggled to transcend the cultural norms and practices of the middle to late nineteenth century well into the twentieth century. Rhetoric and reality sometimes remained quite distinct, as this quick look at Cornell's history reveals.

In many ways, the land-grant idea was a bold experiment. It "transformed higher education through the concept of service and direct links with industry and agriculture . . . and expanded access to higher education" (Altbach, 2011, p. 17). It was an opportunity for greater access as well as the development of scientific research through the increasingly robust models of graduate education (e.g., Shaffer, 2012). In many ways, land-grant colleges and universities began the development of science and technology in the United States that we see as a cornerstone today (Carmichael, 1961, p. 67).

One particularly unique feature of land-grant universities is the existence of agricultural experiment stations that receive considerable federal, state, and private funding for research that informs educational work, both through teaching at the universities and in engagement work on the part of Cooperative Extension. Experiment stations have been and continue to be deeply engaged with the teaching and training of students whose work serves a public purpose, especially scholarship related to agricultural issues. Yet the way this public purpose is defined fits narrowly within the heroic metanarrative embodied in much of the story of what the land-grant institution is and does (Peters, 2006b, 2013, 2017). The service intellectual tradition views one's work through the lens of creating knowledge for others to use (Peters, 2010, pp. 24-32). It also features characteristics of what is often assumed about the work of scholars: that social scientists (or other academic professionals) must maintain a stance of disinterested and unbiased neutrality about their work. It is in this narrative that "historical facts, events, and characters are selected and storied around an ascending plotline of steady progress through science-based service" (Peters, 2013, p. 338).

While these institutions were created to afford opportunities to citizens of lower classes who had previously been unable to attend college, to democratize higher education by opening its doors to those otherwise excluded, and to engage in research with a public purpose, many remained marginalized within these colleges and universities. Some of the most striking examples of discrimination took place in the Southern states. This led to the Second Morrill Act, twenty-eight years after the original creation of the land-grant system.

The Second Morrill Act of 1890 and the Equity in Educational Land-Grant Status Act of 1994

Northern members of Congress passed the First Morrill Act when members from Southern states were absent because of secession and the establishment of the Confederate States of America. After the Civil War and the reintegration of the Confederate states, there was a need to address the reality that Southern states continued to have legalized racial segregation. While some states provided funds for the education of African Americans from the Morrill Act of 1862 at private institutions such as Virginia's Hampton Institute and South Carolina's Claflin University as well as Mississippi's public Alcorn University, the majority of Southern states took no action until they were "induced to do so under the terms of the Second Morrill Act of 1890" (Eddy, 1957, p. 258).

The act stipulated that "no appropriations would go to states that denied admission to the colleges on the basis of race unless they also set up separate but equal facilities" (Rudolph, 1962, p. 254). This legislation provided funds and resources to historically Black colleges and universities (HBCUs), creating educational opportunities for African Americans in Southern states despite a prevailing climate of inequality (Spikes, 1992). However, it should be noted that although most of these institutions were established following the Civil War and before 1900, their growth and development were restricted by lack of financial resources. This was true for general support for these institutions as well as for their explicit land-grant research well into the 1970s when, finally in 1972, 1890 institutions became part of the regular annual appropriation for agricultural research from the U.S. Department of Agriculture (USDA) rather than receiving funds through a special grant renegotiated each year (Christy, Williamson & Williamson, 1992, pp. xvii–xxi).[2]

But even with this annual commitment, there remains chronic underfunding, and recent reports have highlighted the extreme shortfalls for 1890 institutions (Lee & Keys, 2013; Weissman, 2021). As Jim Hightower noted in his report "Hard Tomatoes, Hard Times," the deep, dark secret of the land grant system was the disparity between the 1862 and 1890 institutions: "The black colleges have been less than full partners in the land grant experience. It is a form of institutional racism that the land grant community has not been anxious to discuss" (Hightower, 1978, p. 11). As noted in a 2013 Association of Public and Land-grant Universities (APLU) brief, "the land-grant system is strongest when all universities—1862s, 1890s and 1994s—are funded adequately to carry out the land-grant mission" (Lee & Keys, 2013, p. 2). The fact that state legislatures refused to "adequately support these colleges" and, instead, "direct[ed] larger funds to White universities" highlights the serious challenges 1890s and other HBCUs have always faced (Wade, 2021, p. 10).

B. D. Mayberry noted that the initial and most significant contribution of the 1890 institutions was to provide the mechanism for "4 million negroes (former slaves) to move into the mainstream of American society as citizens with all the rights and privileges embodied in citizenship through education" (Mayberry, 1991, p. 36). Education was and continues to be a central factor in shaping human development. Thus, the creation of these institutions afforded opportunities for African Americans in ways that had been unavailable previously, aside from the private institutions that predated this period.

These institutions remain actively engaged in carrying out the tripartite land-grant mission of teaching, research, and service while maintaining commitments to those disadvantaged by racism and prejudice.[3] With slavery being only a recently abolished practice and Jim Crow shaping much of public life, the 1890 institutions had to address the low educational levels of African Americans. The 1890 institutions began "with elementary and secondary students as the largest portion of their enrollment and worked hard to achieve normal school status" (Humphries, 1991, p. 4). Into the twentieth century, there were few educational opportunities across the board. In 1915, there were "sixty-four public high schools for Blacks in southern states, and only forty-five of them offered a four-year curriculum" (Humphries, 1991, p. 4). As late as 1928, in the seventeen states with Black land-grant institutions, there were 12,922 students enrolled (73 percent) enrolled in private colleges and only 3,527 (27 percent) in Black land-grant colleges.

Particularly for Extension, the necessity for maintaining cultural and political norms led to the formation of separate and unequal structures. For example, in 1915, Texas became the twelfth state to segregate Extension when "[Texas Agricultural Extension Service] administrators created the Negro Division," with the first agents being chosen carefully by officials "believing that they would convey only information that did not threaten white authority"; these first three agents were warned that "if [they] succeeded others would be added to the force and if [they] failed that there would be no other Negro agents employed in the near future" (Reid, 2007, p. 22). Looking at annual reports from several years, one finds that programs and subsequent documentation of Extension's work emphasizes the work with white agents. In the 1936 report from Clemson Agricultural College (now University), the personnel beyond the central administration include "twenty-seven subject matter specialists . . . representing all lines of agriculture in the state. . . . County workers include 46 county agents, one in each county, 15 assistant county agents, and three assistant county agents in soil conservation." Home demonstration was similarly staffed with "the state home demonstration agent, an assistant state agent, and three district agents." There were seven home demonstration specialists and county workers including forty-six home demonstration agents, again one in each county, with three assistant demonstration agents. In contrast, "Negro extension work [was] supervised by a . . . district agent and a . . . supervising agent for home demonstration work." There were sixteen county agricultural agents and fourteen home demonstration agents (Clemson Agricultural College, 1937, p. 2). In the forty-three pages of descriptions of the Extension work taking place across the state, three at the end of the report focused on Negro demonstration work. There were nine modest paragraphs describing the entirety of the work that included field crop production, home improvement and land ownership, 4-H, and food and nutrition (Clemson Agricultural College, 1937, pp. 41–43). This was reflective of the broader culture of Clemson at the time. It was a period in which "administrators relegated Black wage workers to 'Negro houses' that lacked indoor plumbing well into the 1930s; minimized the contributions of talented agricultural agents who labored in the segregated extension service program run by the US Department of Agriculture; hired prominent but 'respectable' Black musicians beginning in 1920 to perform at student-organized segregated social events; and resisted desegregation until 1963" (Thomas, 2020, p. 163). This was for a state population that was 54.3 percent white and 45.6 percent Black in 1930 and 57.1 percent white and 42.9 percent Black in 1940.

As a 1988 USDA report about the 1890s institutions noted, "In the broadest terms, the clientele of the 1890 Extension Programs are people with disadvantages that prevent them from achieving their full potential; disadvantages that turn those people who could be assets to our society into liabilities, unless they are reached with assistance" (U.S. Extension Service, 1988, p. 5). The ability to reach and engage people in rural areas as well as "urban communities—particularly communities in depressed, inner-city areas," has given Extension a significant role to play in a range of settings (U.S. Extension Service, 1988, p. 11).

In a similar spirit, the Equity in Educational Land-Grant Status Act of 1994 provided land-grant designation to thirty-three tribal colleges for Native Americans in Western and Plains states. This provided federal funding for teaching, research, and outreach responding to the specific needs and interests of the Native American populations these institutions serve. Because many of these institutions serve remote populations, the 1994 act provided funds to increase Extension work in areas such as agriculture, community resources and economic development, family development and resource management, 4-H and youth development, leadership and volunteer development, natural resources and environmental management, and nutrition, diet, and health.

What makes these institutions slightly different from their predecessors is the fact that these tribal colleges include community colleges, four-year institutions, and some institutions with graduate-level courses and programs. This group within the land-grant system continues to play an important role in increasing social and economic opportunities for Native Americans through affordable education as well as programs that respond to the particular needs of Native American communities.

While access and affordability have helped to define public higher education, the research and engagement dimensions of the land-grant mission through Cooperative Extension have come to define land-grant colleges and universities because of the commitment to place-based education grounded in relationships with youth through the entirety of the lifespan.

The Hatch Act of 1887, the Smith-Lever Act of 1914, and Earlier Engagement

The ability to realize the public mission of land-grant institutions relies, in large part, on the Hatch and the Smith-Lever Acts. The Hatch Act established the agricultural station system in each of the colleges under the Morrill Act of 1862 to "aid in acquiring and diffusing among the people of the United States useful and practical information on subjects connected with agriculture, and to promote scientific investigation and experiment respecting the principles and applications of agricultural science" (Eddy, 1957, p. 97). This act established and expanded experiment stations across the country on the campuses of land-grant colleges and universities. Some faculty in land-grant colleges of agriculture have appointments that connect their research to experiment station work and include "Hatch" research funds for original research on issues impacting the agricultural industry and rural life (Committee on the Future of the Colleges of Agriculture in the Land Grant System National Research Council 1995, p. 8).

Originally, the Hatch Act had impacts on predominantly white populations as well as for Blacks. The establishment of an experiment station at Tuskegee "[made] it possible for the school to conduct research. Soon afterwards [George Washington] Carver became the director of what became known as the 'movable school,' a stage coach in which lecturers would travel over the county on week-ends to educate Negro farmers on new agricultural approaches based on research conducted on the institutions [*sic*] farm" (Comer et al., 2006).

Five years after the Hatch Act's passage, in February 1892, the first annual Negro Farmers Conference was held, drawing over five hundred farmers to the Tuskegee Institute from across the state: "This conference is said to be the spark that ignited agricultural Extension work among Negroes. The objectives of the movable school were not only to demonstrate new farm practices but also to find out the needs of the farmers and get them the information. The second objective was to get those being educated to use their education in helping the masses" (Comer et al., 2006, para. 18). Thomas Monroe Campbell, the first black Extension agent at Tuskegee, was asked by Booker T. Washington to begin the "new and rather peculiar type of work" then known as the Farmers' Cooperative Demonstration (Denton, 1993, p. 113).

The Smith-Lever Act of 1914 established the Cooperative Extension Service involving the USDA, land-grant colleges and universities, and local communities, with the goal of educating and working with citizens through programmatic initiatives. Often Cooperative Extension shared information with citizens about current developments in agriculture, home economics, and other relevant subjects. But Cooperative Extension also engaged in work with citizens seeking to address challenges facing individuals and communities that went beyond situations that only required technical expertise and knowledge. Cooperative Extension's role goes beyond the application of research-based information to include important community work and leadership development.

Cooperative Extension increased human and monetary capital through public work. But the idea of Extension has roots deeper than the Smith-Lever Act. C. Hartley Grattan notes that "the first quarter-century of land-grant college history was one of toil and struggle, complicated by uncertainty of direction and unclear ideas about what and how to teach the students drawn to the colleges, and how to make the cumulating knowledge available to dirt farmers" (Grattan, 1955, p. 201). Students who attended land-grant colleges went on to become faculty and administrators at these institutions, with Liberty Hyde Bailey being one of the most striking examples of a farmer who was a student at Michigan Agricultural College and then went on to shape both academic life and the lives of many rural people and communities at the turn of the twentieth century as a faculty member and then later as dean of Cornell's College of Agriculture.

Bailey helps us to think critically about the history of land-grant institutions and Cooperative Extension because he saw agricultural education as a means to awaken in rural people a new view of life, rather than simply as a conduit of technical information (Peters, 2006a). He saw colleges of agriculture and experiment stations having important roles in the "future welfare and peace of the people" to a degree that was then yet unforeseen (Bailey, 1915, p. 98). For him, the role of these institutions was in relationship to citizens in order to help them see the world differently and to act differently. Bailey articulates this point:

> The college may be the guiding force, but it should not remove responsibility from the people of the localities, or offer them a kind of co-operation that is only the privilege of partaking in the college enterprises. I fear that some of our so-called co-operation in public work of many kinds is little more than to allow the co-operator to approve what the official administration has done. (Bailey, 1915, p. 100)

Bailey was suspect of what often was identified as engagement with citizens. Land-grant universities had an important role to play, and the faculty within them were important contributors to society, but the ways that university faculty and Extension educators worked with citizens could vary widely. In some situations, if not many, it was appropriate for faculty to utilize their expertise. But one should not confuse the dispensation of facts with education. Peters quotes Bailey from a speech given on December 13, 1899, to the annual Farmers' Convention in Meriden, Connecticut highlighting this point:

> We know that we can point out a dozen things, and sometimes thirteen. But after all, it is not the particular application of science to the farm which is the big thing. The big thing is the point of view. The whole agricultural tone has been raised through these agencies. People are taking broader views of things and of life. Even if we did not have a single fact with which we could answer these people, it is a sufficient answer to say that every agricultural college and every agricultural experiment station, with all their faults, has been a strong factor in the general elevation of agriculture and the common good. The whole attitude has changed. It is the scientific habit of thought and no longer the mere extraneous application of science. (Peters, 2006a, p. 212)

The role of the land-grant university broadly and the experiment station specifically was to do more than provide information. It was also about working with citizens to help realize a different way of seeing the world. This is what Bailey referred to as the "scientific spirit" (Armitage, 2009; Peters, 2007). Defining one's work in such a way challenges a dominant narrative about what land-grant universities were doing during this formative period around the turn of the twentieth century when it seems that what was needed was a "system capable of proving to farmers that 'book farming' was not a joke and that agricultural science, properly applied, would produce a better life for them and their families" (Scott, 1970, p. x). There were competing tensions: one suggested that all we needed were technical skills from scientists at the university, while the other complemented technical knowledge with the essential belief that citizens needed to contribute their own knowledge while also considering new information from research.

In the early 1890s, Pennsylvania State College, Cornell University, and the University of Illinois lent impetus to the Extension concept by adapting techniques of adult education from the Chautauqua movement to engage farmers. By 1907, at least thirty-nine land-grant colleges were "doing *something* in the way of extension" (Grattan, 1955, p. 202). The approaches to programming were diverse. They included lectures, short courses, summer schools, bulletin reports, circulars, cooperative experiments, exhibits at fairs, and demonstrations on farms.

Before and after the Smith-Lever Act, Seaman A. Knapp's demonstration method was foundational to the educational methods of land-grant colleges and Cooperative Extension. While Knapp placed much emphasis on economic gains, he was not solely focused on efficiency and technical expertise. Rather, his ultimate aim was "the development of a vibrant rural civic and cultural life" (Peters, 1998, p. 133).

The language of the Smith-Lever Act reflects Knapp's demonstration model of education in its pronouncement that Extension "shall consist of the giving of instruction and practical demonstrations in agriculture and home economics." This approach was employed in order to "aid in the diffusing among the people of the United States useful and practical information on subjects relating to agriculture and home economics" (Smith & Wilson, 1930, p. 365). Importantly, the language of the Smith-Lever Act was not exclusively aimed at rural people. Rather, it was intended for all people within the United States. Today, much of Cooperative Extension's work takes place within urban and suburban settings addressing and responding to the needs of these communities.

In 1914, there was significant debate and disagreement over exactly "why a national system of agriculture was needed, what it was specifically supposed to accomplish, and how it ought to go about accomplishing it" (Peters, 1998, p. 25). As noted in a 2018 volume on contemporary public engagement work through Extension, "the Smith-Lever Act doesn't say anything at all about purpose" (Peters, Alter & Shaffer, 2018a, p. 24). In short, there was never a unified mission or purpose for land-grant institutions or Cooperative Extension. Since their respective origins, how these institutions should fulfill their public mission has remained in question, although some would argue that the mission has always been a particular way. The dominance of the heroic metanarrative gives the false sense there could not be anything else.

Recognizing this contested beginning for Extension is imperative because the narrative often told about it is that of a single purpose—to use expertise to address issues in communities. Yet, as the chapters of this book highlight, there are significant ways in which expertise is interwoven with a commitment to democratic engagement with diverse individuals and communities:

> To make a long story short, the dominant view of Extension's main purpose, both today and during most of its history, is to address problems in democracy through the provision and application of scientific and technical knowledge and expertise. But here's where Extension becomes interesting. During its long history, many women and men articulated an aspirational vision of Extension's purposes and practices that connected work on problems *in* democracy with problems *of* democracy, in ways that were explicitly attentive to the cultivation of democratic practices. (Peters, Alter & Shaffer, 2018a, p. 24)

The Morrill Acts of 1862 and 1890 and the Equity in Educational Land-Grant Status Act of 1994 have created a system of higher education in all fifty states and several U.S. territories rooted in the understanding that there *was* and *is* a need to have what many refer to as the "People's University" (Sherwood, 2004, p. 2). Additionally, the Hatch Act of 1887 and the Smith-Lever Act of 1914 provided mechanisms necessary to empower the vision of educators to create a system of higher education reaching well beyond the confines of a college campus into communities across the United States. Higher education's role in American democracy goes well beyond the classroom, and Extension has been one of the most important—if not most forgotten—forms of community-based education and development in this country. Yet, as was noted earlier, the ability to leverage organizations such as Extension within diverse communities was impacted by political and cultural limitations based on racism.

To many, land-grant colleges and universities have been "the most celebrated and successful example of the articulation and fulfillment of the service ideal" (Crosson, 1983, p. 22), and the public service mission became institutionalized with the Smith-Lever Act, with the land-grant university continuing to embody that ideal (McDowell, 2001, pp. 15–27). If we look at a historical example from Tuskegee, for example, we see the reach of Extension and its impact: "By 1900, extension through the zealous work of over 1,000 Tuskegee students was well established in twenty-eight states, Cuba, Jamaica, Africa, Puerto Rico, and Barbados" (Denton, 1993, p. 107). But the 1890 Extension system was challenged to provide its services "because of funding disparities between the 1862 and 1890 Extension programs" (Westbrook, 2010, p. 6). This was due, in part, to the fact that "Southern states were not required to match the federal funds provided to the 1890 land-grant institutions" (Westbrook, 2010, p. 6). During the 1930s and 1940s when Extension agents were being trained in democratic discussion methods to facilitate community discussions across the country, the efforts were found across the entire Extension system—within both 1862 and 1890 institutions and their respective educators and communities (Program Study and Discussion Section USDA, 1937; Shaffer, 2017c, 2018b, 2019b). But there were glaring realities of racism built into these efforts, even for efforts advocating for democracy during the rise of fascism and authoritarianism in Europe. The struggle for racial equality was demonstrated by land-grant institutions convening whites and Blacks separately to discussion American democracy and civic life (e.g., Du Bois, 1941).

To understand these historical examples of Extension as well as current models, there is a benefit to looking back at the first decades of Extension's formal existence and the way in which community-based work was taking place and being developed. It is in understanding the shifts about how to develop community-based programs that we can, again, see the importance of cooperative relationships and engagement.

Program Development in Cooperative Extension

Extension was, in the words of Morse Cartwright, executive director of the American Association for Adult Education, "far and away the largest domain of adult education participation" in either the 1920s or 1930s (Cartwright, 1935, p. 119). By the 1940s, Extension was the largest rural adult agency in the world (Brunner & Yang, 1949, p. vii). Extension played a central role in shaping the landscape of adult education as well as that of rural America because of its breadth of programs and sheer presence. The following discussion will identify and address Extension's work during these formative years.

Three Early Phases of Extension Programs

Extension education experienced dramatic changes in its first decades of existence and even prior (Scott, 1970). The changing landscape of Extension's work is captured in a dramatic way in the following account, reinforcing why this period is worthy of study: "The historian of tomorrow will look back on the years 1932–1936 as the most significant epoch in the development of the Agricultural Extension Service since those first 5 years when, before its formative period was over, it faced the emergency demands of the World War" (Brunner & Lorge, 1937, p. 180). The significance of this period, as well as the late 1930s and early 1940s, was primarily based on the shift into new territory where Extension engaged more broadly with communities around issues *beyond* agricultural production.

Smith and Wilson (1930) contended that programs within Extension had gone through three distinct phases of development since its inception in 1914. It should be stated at the onset that while these phases did exist, fluidity also existed in both how programs were developed and how they were implemented. It would be artificial to claim these phases as fixed.

First was the phase "when the government or its extension agent assumed to know what was needed." In this phase, programming came primarily from Extension and was "execution oriented" (Boone & Kincaid, 1966, p. 89). Programs were often predetermined, and farmers received "what was offered by the pioneer agents" (Morris, 1937, p. 2). The demonstration method, pioneered by Seaman A. Knapp, was the preferred method, especially since thousands of farmers could readily see demonstrations and, in turn, improve farm practices.[4] The demonstration method also shaped work in other areas such as horticulture (L. G. Smith, 1938). The language of the Smith-Lever Act reflects the importance of the demonstration method stating that Extension "shall consist of the giving of instruction and practical demonstrations in agriculture and home economics" (Smith & Wilson, 1930, p. 365). But quickly Extension leaders recognized that they could not only provide information to farmers.

The second phase shifted away from the previous model because farming people, with Extension in consultation with them, were now largely responsible for program development. Programs were "self-determined" and largely based on local conditions and information. As Morris noted, "This was a very significant period in the history of extension program planning. Thousands of farmers, for the first time in their experience, were given an opportunity by the extension service to gather around farm dining-room and kitchen tables, in schoolhouses and grange halls, collectively to study and plan action on their problems" (Morris, 1937, p. 3). The coming together of community members to cooperate in program planning was a significant shift from models where experts produced programs and then took them to individuals or communities. In many ways, Extension played an important role in community development by developing programs "*with* people rather than *for* people" (Phifer, List & Faulkner, 1980, p. 21).

We see the importance of cooperation in M. C. Burritt's telling of how to build a program: "Can there be any better means than the coming together of representative committees and councils of farmers from the communities to consider their needs and to determine upon the best ways to meet them? The people who have lived long in a community have learned its handicaps and its limitations as well as its advantages" (Burritt, 1922, p. 8). And while farmers and other rural people came together to lay out their interests and problems, it became apparent that planning programs in this "free grassroots approach" was not sustainable with limited personnel (Boone & Kincaid, 1966, p. 90). Local leadership was not developed to a degree that enabled such a variety of programs to be supported. A new approach was needed, and this led to yet another phase.

The third phase was an adjustment of this relationship because Extension agents and farming people "together made the analysis of conditions, together selected the outstanding needs, and together made a program to meet those needs" (Smith & Wilson, 1930, p. 132). This was a new role for Extension agents in that they would "guide and help to develop people rather than tell them" what to do (C. B. Smith, 1938, p. 177). This challenge would be present when one looks at the discussion movement of the 1930s and early 1940s through Extension. Agents were resistant to elements of discussion groups in communities and implementing these approaches in addition to more technical work because agents "were accustomed to parceling out a continuous supply of 'right answers' to immediately pressing farm problems and consequently often found it difficult to see the practical value of philosophical discussion groups." Gladys Baker, in her book *The County Agent*, continued: "Leaders in some states reported that it was difficult to keep the county agent from monopolizing the discussion and insisting that his viewpoint and judgment be accepted by the group" (Baker, 1939, p. 85).[5] Ideally, this phase brought together the knowledge of various experts as well as farmers to create programs. It is important to note that women were becoming more engaged in program planning, moving away from limited roles exclusively for programs in home economics and instead participating alongside men in all aspects of program planning for farm life (Peterson, 1938).

Increasingly, this third phase became the norm because of the ability to connect local Extension programs to other initiatives taking place elsewhere—regionally, nationally, or globally. Yet while participation on the part of farmers was increasing, *who* those farmers were played an important role in shaping programs. Those involved with program planning at the community level were often those with political agency beyond the typical rural farming family. County planning committees of "3 to 12 or 15 committeemen" did not give adequate representation for those from the community (Morris, 1937, p. 8). The voice of lower income farmers and their families would often struggle to be heard among those with vested interests in agriculture. This was especially true for Black farmers (Wood, 2006).

State Extension programs, by and large, dealt with broad objectives that would ideally guide community program making (Smith & Wilson, 1930, p. 139). For most issues, similar to local situations, special committees composed of bankers, businesspeople, representatives of farmers' organizations, the press, and outstanding farmers would gather at the land-grant college and participate in statewide agricultural conferences. Additionally, faculty members with specialties within a given area of interest would share "the facts relating to the commodity already assembled by the college" (Smith & Wilson, 1930, p. 139). Each committee would report to the state conference, and what they would approve became the state agricultural program for Extension work. The recommendations made by these statewide committees through the state conference would then be taken to each county within the state and to representatives from various stakeholder perspectives. County committees would then ensure that county programs would meet local needs while aligning with the state program.

Another issue that emerged at the state level of program planning was that the content of Extension programs and the methods by which they were determined reflected the "philosophy of Extension held by those responsible for leadership" (Brunner & Yang, 1949, p. 103). There were differing attitudes within Extension regarding program planning. A tension existed between having predetermined programs created by the state office and the approach of local groups having sole responsibility for planning and developing projects (Brunner & Yang, 1949, p. 103). In a study of Extension work in Iowa, we see a perspective about programs being coordinated by Extension professionals without participation from citizens.

> It is necessary in extension teaching to plan carefully every phase of work to be offered in the field because it is from the plans that programs are selected in the counties. Each phase offered is written into an outline known as the "project." When work is to be offered in the field the usual practice is for the specialist to prepare a project outline covering the objectives, the methods to be followed and the results that may be expected. . . . A complete set of these project outlines for all lines of work is then prepared and used by county and local committees for program making in the fall of each year. The combined programs of the counties constitute the state extension program, and each specialist thus has an annual program mapped out before the beginning of each year. (Davidson, Hamlin & Taff, 1933, pp. 93–94)

Here we see Extension having the program objectives, methods, and expected results before ever stepping into a community. In this example, the work of Extension is easily organized. The year is planned, barring any emergencies or unforeseen circumstances. While this is not a reflection of all Extension programs, it does highlight the continuation of earlier models when programs were created without including citizens in the planning processes.

Responding to such views of Extension's programmatic work, the *Joint Committee Report on Extension Program, Policies and Goals*, an effort between USDA and the Association of Land-Grant College and Universities, called attention to the reality that many of Extension's programs and procedures were developed prior to and during World War I and that there was a need to study and reevaluate their work. The *Joint Committee Report* also echoed the sentiment that Extension programs were "based largely on the agents' analysis," and despite attempts to democratize the formation of programs, the tendency was for the state offices to formulate a program and "make them like it." The report acknowledged some of the other criticisms of Extension by noting that the planning process was one of "form rather than substance" and that too often "sizable groups" of especially lower income farmers as well as other marginalized groups were not taken into consideration in program planning. Finally, the report concluded that Extension had been faulted for focusing on issues related to agricultural production to the exclusion and detriment of "many other problems and issues of vital concern" (Hannah et al., 1948, p. 36). To some, Extension had a choice between adult education or a more focused technical mission (Miller, 1973, p. 18). For others, the issue boiled down to whether the job of Extension was primarily vocational and thus should view citizens as agents of production, or whether it should consider the total needs of citizens and view them as human beings (Brunner & Yang, 1949, p. 104). To highlight this point, it is worth exploring one of the important voices from these first decades: Home Demonstration Leader Minnie Price.

Are the Fundamental Democratic Objectives Education or Training?

Employed by The Ohio State University as Extension home demonstration leader for Ohio State University from 1923 to 1951, Minnie Price was a 1911 graduate of Oregon State Agricultural College. She earned a second bachelor's degree in 1915 from Columbia University's Teachers College in New York City. She had a land-grant education and was immersed in the discussions about the role of education in democracy while at Columbia. In a speech she gave in 1938 at the annual Minnesota Agricultural Extension Conference in St. Paul, she spoke at length about Extension's responsibilities for and contributions to the task of making democracy work as it should.[6]

In her speech, Price sounded a note of urgency. She quoted John Dewey as having said that "the fundamental beliefs and practices of democracy are now challenged as they have not been since the rise of democratic institutions." She went on to say that if Dewey's claim was true, "we might well ask the question, do we in Extension provide opportunity for practice in the democratic way of life?" (Price, 1938, p. 26). Her answer, in short, was "Not enough." In essence, her speech was an invitation to her Extension colleagues to see themselves and their institution as agents in the ongoing story about the building and realization of such a life.

In her speech, which was titled "The Fundamental Objectives of Agricultural Extension Work," Price (1938, p. 28) asked an important question: "Is Extension work education and/or training?" Her question was provoked in part by the memory of a class she had attended several years earlier, during which she heard someone refer to "Extension folks" as "animal trainers." Contradicting this view, she marshaled evidence that Extension leaders understood Extension work to be education rather than training. But she wondered if they really meant what they said. Or, to put it another way, whether what they meant matched what she understood education to be and require.

Acknowledging both that meanings do not remain constant and that people have the right to their own interpretations, Price spent a good deal of the first half of her speech articulating her own perspective on the meaning and purposes of education in a democratic society. In doing so she argued that education "must help life to be lived on an increasingly high level" (Price, 1938, p. 20). Here she was not mainly or only referring to economic and material opportunities and standards. She was referring to social, cultural, and political opportunities and standards that flow from an understanding of democracy as not only a system of government but also a way of life. "Education in a democracy dare not be purely technical," she warned. "It must lead on to the development of inner resources, to an appreciation of the things that enrich life, and to intelligent participation in those movements which govern the type of life possible" (Price, 1938, p. 27).

It was in relation to the ideal of democracy as a way of life that Price located what she referred to in her title as the "fundamental" objectives of agricultural Extension work. Such objectives, she said, "grow out of life and are worded in terms of one's philosophy of life" (Price, 1938, p. 19). In her elaboration of elements that she felt were included in Americans' view of "the good life," she included the following:

> An opportunity for personalities to grow within the limit of their capacities. This is basic to democracy. The democratic ideal emphasizes above all else the importance of persons. Persons are always "ends" and never "means." This democratic concept assumes the right of the individual to attain his fullest

> development and to participate in carrying responsibilities and in managing affairs related to the collective aspects of life. . . . The concept of democracy prescribes participation. (Price, 1938, p. 24)

It was in a section of her speech labeled "Democratic Procedure" that Price included the quote from John Dewey—whose class on the philosophy of education she had most likely taken while she was a student at Teachers College—about the beliefs and practices of democracy being challenged as never before. After quoting Dewey she followed her question about whether Extension provided "opportunity for practice in the democratic way of life" by saying in no uncertain terms that it should. The reasons why included the value of such opportunities for the work of advancing individual human development. But they also included their value for the work of understanding and addressing both technical and social problems. Importantly, Price argued that when this second kind of work was approached as an opportunity for practice in the democratic way of life, it promised to produce not only sound and effective policies to deal with problems communities and the nation faced, but also better Extension work. "The injection of the thinking and opinions of vast numbers of intelligent thoughtful citizens," she said, "can do much to give in the long run, which is our responsibility, better national policies, laws, education, and even better agricultural Extension work" (Price, 1938, p. 26). To support her perspective she quoted Henry A. Wallace, President Franklin D. Roosevelt's secretary of agriculture. "The strength of the Extension service," Wallace wrote,

> is in its capacity to think through with the members of farm families and with one another the problems of rural life. . . . I suggest that they vigorously seek the truth in terms of advantages and disadvantages, and in this way draw intelligent conclusions. I hold that these conclusions should not be crystalized, but remain fluid and flexible in the face of constantly changing economic conditions . . . leaving to the farmers the course of action to be taken after obtaining a clear conception of the situation. (Price, 1938, pp. 26–27)

Price notes that while Wallace's argument was focused on agricultural policies, "it applies elsewhere." By way of illustration, she quoted the report of President Roosevelt's advisory committee on education:

> The committee is impressed with the obvious fact that few social problems can be solved by the federal government alone. In all the social services, and particularly in education, a high degree of intelligent local initiative is essential. A major endeavor of all national action in those fields should be to foster and preserve the strength of local democratic action. (Price, 1938, p. 27)

How well was Extension doing with this endeavor? In Price's view, not well. "C. C. Taylor said several years ago," she wrote, "that Extension folks had not shown a faith in the knowledge and the capacity for leadership among rural people—and had therefore not even attempted to use such leadership. We all know of places where we have used it, but his charge implies that by and large we have failed." Reminding her audience that education should result "in intelligent participation by individuals in the management of conditions in which they live," Price argued that there

> should be constantly increasing ability on the part of a constantly increasing number of folks to do this. And Extension staff members, unless they assume the role of dictators, must work with rural folks to this end. To the extent that we treasure more and more the democratic ideal, will we provide experiences in practice of initiative, in the carrying of responsibilities, in the making of decisions, in the formation and execution of plans, and in the evaluation of outcomes. Such a procedure, along with the technical education which we have long stressed, could create a worthy farm population, technically equipped and with greater insight into the great flow of political and economic life. (Price, 1938, p. 27)

The invitation that Price and many others at the time offered to Extension professionals was to find themselves a part of a prophetic story about the building of a democratic society. This wasn't just a rhetorical device or a hypothetical argument. It was taken seriously, and its implications with respect to professional routines, purpose, and identity were operationalized across the nation in small and large ways during Extension's founding decades as demonstrated through the discussion efforts of the 1930s and 1940s mentioned earlier. Price offers a reminder that Extension has always wrestled with questions of identity and purpose, exploring nuanced responses to the question, What is education? In a natural way, this question of education leads to the programmatic efforts to make that possible through planning. Price's words animate, in certain situations, the practical planning and development of programs with communities. But how was planning approached, more broadly?

Two Studies of Program Planning

In one of the few studies focused on program planning from the period, Fred Morris (1937) laid out a framework for thinking about program planning. Morris identified necessary steps to program planning and also some serious weaknesses that should have been addressed to ensure that program planning would be more successful and participatory.

First, he argued that one must define the chief function of Extension. In the deepest sense, Extension work was designed to develop the person as a member of society. By defining Extension work as something that is primarily concerned with the development of individuals—rather than products—it served as a reminder that human capacity for change took precedent over improved harvests.

Second, there should be formulated in clear and understandable language the general aims of Extension education. Such an aim could include goals and objectives. Once the aim was determined, the next question in need of clarification was, What is the county agricultural program? In Morris's study, the review of programs from counties in fourteen states failed to give a satisfactory answer. In general, programs were lists of projects. Finally, Morris (1937, pp. 11–12) laid out procedures for how county program planning should happen:

1. Clarify aims and objectives of extension education.
2. Define county agricultural extension programs and plans of work.
3. Provide farmers with ample opportunity for continuous and genuine participation in planning.

4. Get specialists, county agents, and farmers to approach program planning by the collection, organization, analysis, and interpretation of county situations, and by searching for larger problems.
5. Provide for coordination of plans of work for the basic problems.
6. Give attention to the closer correlation of extension and research.

The two "glaring weaknesses" Morris called attention to were the lack of effective farmer participation in planning and the failure to get sufficient community and county farm data as a basis for problem determination. Without local people participating as fully as possible and sharing their local knowledge about the land and environment, the give-and-take process of Extension work was diminished or lost.

Brunner and Yang (1949, pp. 106–110) also offered an assessment of what program planning for Extension should include based on their study of Extension's then thirty-five-year history. Their principles were:

1. Programs should be planned with and not for the people concerned.
2. Local leadership is indispensable to the development and execution of an Extension program.
3. Any educational program must be based upon facts.
4. Programs should not only be useful, but also of high interest in terms of a recognized problem.
5. For programs to be effective, the process requires time.
6. When teaching farmers new things, one should always hold the big idea foremost in their attention.
7. Use local material for planning and development of an Extension program.
8. Any educational program must take the habits, customs, and culture of the people into account.

These two studies both emphasized the importance of participation in planning and the need for long-term investment in such processes. While participation did occur, there was enough of a continuation of a top-down, expert-dominated model for program development that such critiques were warranted. Importantly, these issues have long continued to be factors regarding the participation of citizens in program planning (Cervero & Wilson, 2006).

Methods of Implementation and Philosophies of Education

The subjects and issues addressed through programming were continuously expanding as were the techniques. For more than three decades, Extension developed and tested different teaching methods adaptable to the circumstances it faced (Hannah et al., 1948, pp. 33–34). Nevertheless, demonstration was the "foundation stone" for Extension teaching and remained a dominant method (Brunner & Yang, 1949, p. 113). In 1931, for example, more than 1.1 million local demonstrations took place on farms for the improvement of practices in agriculture and home economics (Landis, 1934, p. 1). But this was not to the exclusion of other methods during this period. Educators used demonstrations, exhibits, photographs, lantern slides, motion pictures, posters, charts, filmstrips, television, farm and home visits, office and telephone calls, meetings, radio on the farm or in rural communities, bulletins and circulars, and news stories (Brunner & Yang, 1949, pp. 112–132; Davidson, Hamlin & Taff, 1933, pp. 186–190; Hannah et al., 1948, pp. 33–35; Smith & Wilson, 1930, pp. 271–289). But in addition to these various methods of communication, there were efforts to deeply intertwine the programs with a sense of its purpose for democracy and the role of people at the heart of that practice. For example, M. L. Wilson of the USDA wrote about discussion as the "archstone of democracy" in the *Extension Service Review* in reference to efforts by Extension agents to infuse citizen voices into community-based work (Wilson, 1935). The ideals of democracy were deeply practical, especially in communities across the country as people collaborated with the local Extension office.

Assumptions about what Extension was, and is, become important as we think about its audiences and the role it plays. As recounted in a 2003 volume about the work of Cornell University Cooperative Extension-NYC by Madie McLean, a community nutrition educator with then more than thirty years in Extension, the work of Extension is more complex and rooted in a sense of opportunity and justice for people, especially those on the margins of society:

> After telling us a bit about her life and how she came to work with extension, she launched into a story about a young woman in one of her basic nutrition classes who didn't know how to cook rice. She recounted how they practiced together—two cups water, one cup rice; two cups water, one cup rice. The following week, the young woman came back to class beaming with a new-found pride and confidence; she had made rice for her family's dinner. At this point, Madie stopped and looked at us pointedly: "If you want to understand what we do here," she said, "you have to understand, it's not about the rice." (Hittleman & Peters, 2003, p. 6)

Since its earliest days of dealing with the boll weevil or teaching people how to can produce or how to become active and engaged citizens through 4-H, Extension has taught people how to "cook the rice," but also created the conditions to have a learning opportunity that is both technical and civic, informative and explorative. In addition to providing answers and solutions, Extension has also played a role in helping to develop relationships and cultivate what Ruby Green Smith called "the good life" (Smith, 1949, p. 544).

If we jump forward to current discussions about the land-grant civic-engagement mission, we find a familiar discourse regarding new practices and approaches with respect to how to transition from service and outreach to engagement. With roots in communities through teaching and research, land-grants were, by and large, transformed after World War II from locally led programs to more research-based efforts. The establishment of the National Science Foundation helped to push forward the research university agenda. Community-based educational opportunities became secondary to the investment in basic research aimed at solving the world's problems (Vest, 2007, p. 24). The Kellogg Commission on the Future of State and Land-Grant Universities (1999, p. 27), whose report has been central to the engagement mission of the modern university, noted that "it is time to go beyond outreach and service to what the Kellogg Commission now defines as 'engagement.'" What's striking is that some of the "new scholarship" articulated by the likes of Boyer (1990, 1996) and Schön (1995) highlighted the importance of not viewing the work of the university exclusively through the lens of technical rationality. As Fear and Sandmann (2001–2002, p. 30) note, "Schön contended that the 'new scholarship' conflicts philosophically with the prevailing ethos of the research university, academe's most prestigious and admired institutional expression." This shift, as expressed by the Kellogg Commission report, to move from outreach and service involving "one-way transfers of university expertise" to "a two-way relationship between higher education and society" with a commitment to collaboration, was rooted in a commitment to broadening the epistemology that has animated the research-intensive university since the middle of the twentieth century (Glass & Fitzgerald, 2010, p. 13).

The engagement movement of recent decades has raised important concerns about knowledge and what it means when we recognize that we have systemically privileged one institutional epistemology while discounting others (Hall & Tandon, 2020; Shaffer, 2017b). But as we can see from this brief look at the roots of the land-grant university and of Extension's early history, the relational and collaborative model of what Peters (2015, p. 45) describes as "political in a local community- and neighborhood-centered sense" informs how we think about the commitment to using science and expert knowledge alongside lived experience and common sense to address shared wicked problems (e.g., Fischer, 2000). As scholars continue to push our thinking about who has knowledge and who does not, we can learn from the first years of Extension as a time prior to our familiar research-centric model that so much of the modern engagement-movement challenges with respect to the dominant research-centric model's influence on our universities.

There are many examples from Extension's long and rich history to offer glimpses into Extension's narrative that is first and foremost about building a life and culture for people. We conclude with two statements from prominent leaders of Extension. The first comes from the director of Extension work, Clyde William Warburton, in a 1930 article:

> For what is the object of extension work? More bushels of corn? More bales of cotton? More pounds of butter fat in the dairy cow's annual record? More quarts of fruit and vegetables canned for winter use? No, these are but means to an end. The end, the object of extension work, is to aid the farmer and his family to improve living conditions on the farm, to provide a more satisfying rural life. . . . Better crops, better livestock, better food, better clothes, these are among the objects of extension work. But back of it all, the ultimate purpose is to create better homes, better citizens, better communities, better rural living. (Warburton, 1930, pp. 292–293)

Warburton articulated a belief that Extension was more than improving yields for farmers. But determining how to accomplish the goal of creating better citizens and communities was and is the challenge. Can we make distinctions between the economic and tangible benefits and the civic benefits? The second statement comes in the first paragraph of *The Agricultural Extension System of the United States*, a book also published that same year. It was authored by two national USDA Extension administrators, C. B. Smith and M. C. Wilson:

> There is a new leaven at work in rural America. It is stimulating to better endeavor in farming and home making, bringing rural people together in groups for social intercourse and study, solving community and neighborhood problems, fostering better relations and common endeavor between town and country, bringing recreation, debate, pageantry, the drama and art into the rural community, developing cooperation and enriching the life and broadening the vision of rural men and women. This new leaven is the cooperative extension work of the state agricultural colleges and the federal Department of Agriculture, which is being carried on in cooperation with the counties and rural people throughout the United States. (Smith & Wilson, 1930, p. 1)

Warburton's idea of creating better citizens and better communities alongside Smith and Wilson's notion of Extension serving as a leaven bringing about greater cooperation demonstrate the commitment to doing more than purely technical work regarding agricultural and economic issues. Smith and Wilson (1930, p. 23) touched on this theme of cooperation later in their book when they noted that "it is only when the federal Department of Agriculture, the state agricultural college and the local government and people all unite on a common plan and purpose that the best results are obtained." Yet, as we see from the racial barriers within society and Extension, more specifically, the sense of that commonality and cooperation only extended so far.

The first decades of Extension's history, built upon the foundation of the 1862 and 1890 Morrill Acts as well as the Hatch and Smith-Lever Acts, serve as a reminder of where Extension has been and how, in many ways, we find ourselves dealing with similar questions. As this chapter has highlighted, understanding the elements that have formed Extension historically also provide an opportunity not only to reflect on the legislative acts that brought forth Extension, but also to reconsider the ways in which we think about the development of programs and the underlying purpose behind them—attending to questions about who programs are *for*, *with*, and *why we do them*—today.

NOTES

Portions of this chapter were originally published as Timothy J. Shaffer, "The Land Grant System and Graduate Education: Reclaiming a Narrative of Engagement," in *Collaborative Futures: Critical Reflections on Publicly Active Graduate Education*, edited by Amanda Gilvin, Georgia M. Roberts, and Craig Martin (Syracuse, NY: Graduate School Press of Syracuse University, 2012), 49–74.

1. It should be noted, however, that Morrill was not the first with the idea of public higher education. Jonathan Baldwin Turner suggested that federal land grants be given to states in order to establish industrial universities in the early 1850s. Turner championed this idea for the establishment of a state industrial university in Illinois. See Nevins, 1962.
2. For information about the history of research at 1890 institutions, see Mayberry & Bicentennial Committee of Association of Research Coordinators, 1976.
3. There are nineteen official 1890 land-grant institutions. The then Tuskegee Institute was created by an act of the Alabama legislature in 1881 only to have the state establish and incorporate a board of trustees and name the school private. Nevertheless, it was granted twenty-five thousand acres of land by the U.S.

Congress in 1899 and is a cooperating partner with Auburn University and Alabama A & M University with respect to Cooperative Extension work. For these reasons, Tuskegee is included in lists of 1890 institutions. Of note is Central State University in Ohio, receiving 1890 land-grant status in 2014.

4. While Knapp placed much emphasis on economic gains, he was not solely focused on efficiency and technical expertise. Rather, his ultimate aim was for "the development of a vibrant rural civic and cultural life" (Peters, 1998, p. 133). Categorizing Extension work to "phases" has the potential to flatten the persons or concepts too easily; yet such categorization gives a useful framework for thinking about the changes that took place in the early years of Extension.
5. For more on the discussion movement and Extension's role, see Shaffer, 2017a, 2017c, 2018a, 2018b, 2019a, 2019b.
6. For contemporary examples of such democratic work in Extension, see Peters, Alter & Shaffer, 2018b.

REFERENCES

Altbach, P. G. (2011). Patterns of Higher Education Development. In P. G. Altbach, P. J. Gumport & R. O. Berdahl (Eds.), *American Higher Education in the Twenty-First Century: Social, Political, and Economic Challenges* (3rd ed., pp. 15–36). Baltimore: Johns Hopkins University Press.

Armitage, K. C. (2009). "The Science-Spirit in a Democracy: Liberty Hyde Bailey, Nature Study, and the Democratic Impulse of Progressive Conservation". In M. Egan & J. Crane (Eds.), *Natural Protest: Essays on the History of American Environmentalism* (pp. 89–116). New York: Routledge.

Bailey, L. H. (1915). *The Holy Earth.* New York: C. Scribner's Sons.

Baker, G. L. (1939). *The County Agent.* Chicago: University of Chicago Press.

Boone, E. J., & Kincaid, J. . (1966). Historical Perspective of the Programming Function. In H. C. Sanders (Ed.), *The Cooperative Extension Service* (pp. 89–93). Englewood Clifs, NJ: Prentice-Hall.

Boyer, E. L. (1990). *Scholarship Reconsidered: Priorities of the Professoriate.* Princeton, NJ: Carnegie Foundation for the Advancement of Teaching.

———. (1996). The Scholarship of Engagement. *Journal of Public Service and Outreach,* 1(1), 11–20.

Brunner, E., & Lorge, I. (1937). *Rural Trends in Depression Years: A Survey of Village-Centered Agricultural Communities, 1930–1936.* New York: Columbia University Press.

Brunner, E., & Yang, E. H. P. (1949). *Rural America and the Extension Service: A History and Critique of the Cooperative Agricultural and Home Economics Extension Service.* New York: Bureau of Publications, Teachers College.

Brunner, H. S. (1962). *Land-Grant Colleges and Universities, 1862–1962.* Washington, DC: United States Government Printing Office.

Burritt, M. C. (1922). *The County Agent and the Farm Bureau.* New York: Harcourt, Brace.

Carmichael, O. C. (1961). *Graduate Education: A Critique and a Program.* New York: Harper & Brothers.

Cartwright, M. A. (1935). *Ten Years of Adult Education: A Report on a Decade of Progress in the American Movement.* New York: Macmillan.

Cervero, R. M., & Wilson, A. L. (2006). *Working the Planning Table: Negotiating Democratically for Adult, Continuing, and Workplace Education.* San Francisco: Jossey-Bass.

Christy, R. D., Williamson, L., & Williamson, H. (1992). Introduction. A Century of Service: The Past, Present, and Future Roles of 1890 Land-Grant Colleges and Institutions. In R. D. Christy & L.

Williamson (Eds.), *A Century of Service: Land-Grant Colleges and Universities, 1890–1990* (pp. xiii–xxv). New Brunswick, NJ: Transaction Publishers.

Clemson Agricultural College. (1937). *Extension Work in South Carolina 1936*. Clemson, SC: Clemson Agricultural College.

Clute, O. (1891). The State Agricultural College. In Andrew C. McLaughlin, *History of Higher Education in Michigan*, Bureau of Education Circular of Information 4 (pp. 105–115). Washington, DC: Government Printing Office.

Cohen, A. M., & Kisker, C. B. (2010). *The Shaping of American Higher Education: Emergence and Growth of the Contemporary System* (2nd ed.). San Francisco: Jossey-Bass.

Comer, M. M., Campbell, T., Edwards, K., & Hillison, J. (2006). Cooperative Extension and the 1890 Land-Grant Institution: The Real Story. *Journal of Extension*, 44(3), 3FEA4. https://archives.joe.org/joe/2006june/a4.php.

Committee on the Future of the Colleges of Agriculture in the Land Grant System National Research Council (1995). *Colleges of Agriculture at the Land Grant Universities: A Profile*. Washington, DC: National Academy Press.

Crosson, P. H. (1983). *Public Service in Higher Education: Practices and Priorities*. Washington, DC: Association for the Study of Higher Education.

Davidson, J. B., Hamlin, H. M., & Taff, P. C. (1933). *A Study of the Extension Service in Agriculture and Home Economics in Iowa*. Ames: Iowa State College, Collegiate Press.

Denton, V. L. (1993). *Booker T. Washington and the Adult Education Movement*. Gainesville: University Press of Florida.

Dorn, C. (2017). *For the Common Good: A New History of Higher Education in America*. Ithaca, NY: Cornell University Press.

Du Bois, W. E. B. (1941). As the Crow Flies. *New York Amsterdam Star-News*, February 8.

Eddy, E. D. (1957). *Colleges for Our Land and Time: The Land-Grant Idea in American Education*. New York: Harper.

Farrand, L. (1929). Letter to R. C. Peyton. July 22. In Livingston Farrand Papers, Division of Rare and Manuscript Collections, Cornell University Library. #3-5-7.

Fear, F. A., & Sandmann, L. R. (2001–2002). The "New" Scholarship: Implications for Engagement and Extension. *Journal of Higher Education Outreach and Engagement*, 7(1 & 2), 29–39.

Fischer, F. (2000). *Citizens, Experts, and the Environment: The Politics of Local Knowledge*. Durham, NC: Duke University Press.

Geiger, R. L. (2000a). Introduction: New Themes in the History of Nineteenth-Century Colleges. In R. L. Geiger (Ed.), *The American College in the Nineteenth Century* (pp. 1–36). Nashville: Vanderbilt University Press.

———. (Ed.) (2000b). *The American College in the Nineteenth Century*. Nashville: Vanderbilt University Press.

Glass, C. R., & Fitzgerald, H. E. (2010). Engaged Scholarship: Historical Roots, Contemporary Challenges. In H. E. Fitzgerald, C. Burack & S. D. Seifer (Eds.), *Handbook of Engaged Scholarship: Contemporary Landscapes, Future Directions* (Vol. 1: *Institutional Change*, pp. 9–24). East Lansing: Michigan State University Press.

Grattan, C. H. (1955). *In Quest of Knowledge: A Historical Perspective on Adult Education.* New York: Association Press.

Hall, B., & Tandon, R. (2020). Editorial: Knowledge Democracy for a Transforming World. *Gateways: International Journalof Community Research and Engagement,* 13(1), 1–5.

Hannah, J. A., et al. (1948). *Joint Committee Report on Extension Programs, Policies and Goals.* Washington, DC: U.S. Government Printing Office.

Hightower, J. (1978). *Hard Tomatoes, Hard Times: The Original Hightower Report, Unexpurgated, of the Agribuiness Accountability Project on the Failure of America's Land Grant College Complex and Selected Additional Views of the Problems and Prospects of American Agriculture in the Late Seventies.* Cambridge, MA: Schenkman Publishing Company.

Hittleman, M., & Peters, S. J. (2003). It's Not About the Rice: Naming the Work of Extension Education. In S. J. Peters & M. Hittleman (Eds.), *We Grow People: Profiles of Extension Educators* (pp. 5–13). Ithaca, NY: Cornell University.

Humphries, F. (1991). 1890 Land-Grant Institutions: Their Struggle for Survival and Equality. *Agricultural History,* 65(2), 3–11.

Kellogg Commission on the Future of State and Land-Grant Universities. (1999). *Returning to Our Roots: The Engaged Institution, Third Report.* Washington, DC: Association of Public and Land-Grant Universities. https://www.aplu.org/library/returning-to-our-roots-the-engaged-institution/file.

Landis, B. Y. (1934). Agricultural Extension. In D. Rowden (Ed.), *Handbook of Adult Education in the United States* (pp. 1–15). New York: American Association for Adult Education.

Lee, J. M., & Keys, S. W. (2013). *Land-Grant but Unequal: State One-to-One Match Funding for 1890 Land-Grant Universities.* Washington, DC: Association of Public and Land-Grant Universities. https://www.aplu.org/library/land-grant-but-unequal-state-one-to-one-match-funding-for-1890-land-grant-universities/file.

Lee, R., & Ahtone, T. (2020). Land-Grab Universities: Expropriated Indigenous Land Is the Foundation of the Land-Grant University System. *High Country News,* 52(4), 32–45. https://www.hcn.org/issues/52.4/indigenous-affairs-education-land-grab-universities.

Levine, P. (2014). Civic Studies. *Philosophy & Public Policy Quarterly,* 32(1), 29–33.

Mayberry, B. D. (1991). *A Century of Agriculture in the 1890 Land-Grant Institutions and Tuskegee University, 1890–1990.* New York: Vantage Press.

Mayberry, B. D., & Bicentennial Committee of Association of Research Coordinators (1976). *Development of Research at Historically Black Land-Grant Institutions.* N.p.

McDowell, G. R. (2001). *Land-Grant Universities and Extension into the 21st Century: Renegotiating or Abandoning a Social Contract.* Ames: Iowa State University Press.

Miller, P. A. (1973). *The Cooperative Extension Service: Paradoxical Servant—The Rural Precedent in Continuing Education.* Syracuse, NY: Syracuse University, Publications in Continuing Education.

Morrill, J. S. (1888). *State Aid to Land Grant Colleges, an Address.* Burlington, VT: Free Press Association.

Morris, F. B. (1937) Planning Agricultural Extension Programs. In *Extension Service Circular 260.* Washington, DC: United States Department of Agriculture.

Nevins, A. (1962). *The State Universities and Democracy.* Urbana: University of Illinois Press.

Peters, S. J. (1998). Extension Work as Public Work: Reconsidering Cooperative Extension's Civic Mission. PhD dissertation, University of Minnesota, Minneapolis.

———. (2006a). "Every Farmer Should Be Awakened": Liberty Hyde Bailey's Vision of Agricultural Extension Work. *Agricultural History*, 80(2), 190–219.

———. (2006b). It's Not Just Providing Information: Perspectives on the Purposes and Significance of Extension Work. In S. J. Peters, D. J. O'Connell, T. R. Alter & A. L. H. Jack (Eds.), *Catalyzing Change: Profiles of Cornell Cooperative Extension Educators from Greene, Tompkins, and Erie Counties, New York* (pp. 13–32). Ithaca, NY: Cornell University.

———. (2007). "Laying the Very Foundations of Democracy": The Purpose and Significance of the Spread of the Scientific Spirit in Liberty Hyde Bailey's View of the Civic Ends and Means of Agricultural Extension Work. Paper presented at the Agricultural History Society Annual Meeting. Ames, IA.

———. (2010). *Democracy and Higher Education: Traditions and Stories of Civic Engagement.* East Lansing, MI: Michigan State University Press.

———. (2013). Storying and Restorying the Land-Grant System. In R. L. Geiger & N. M. Sorber (Eds.), *The Land-Grant Colleges and the Reshaping of American Higher Education* (pp. 335–353). New Brunswick, NJ: Transaction Publishers.

———. (2015). A Democracy's College Tradition. In H. C. Boyte (Ed.), *Democracy's Education: Public Work, Citizenship, and the Future of Colleges and Universities* (pp. 44–52). Nashville: Vanderbilt University Press.

———. (2017). Recovering a Forgotten Lineage of Democratic Engagement: Agricultural and Extension Programs in the United States. In C. Dolgon, T. D. Mitchell & T. K. Eatman (Eds.), *The Cambridge Handbook of Service Learning and Community Engagement* (pp. 71–80). New York: Cambridge University Press.

Peters, S. J., Alter, T. R., & Shaffer, T. J. (2018a). Introduction: Making Democracy Work as It Should. In S. J. Peters, T. R. Alter & T. J. Shaffer (Eds.), *Jumping into Civic Life: Stories of Public Work from Extension Professionals* (pp. 13–36). Dayton, OH: Kettering Foundation Press.

———. (Eds.). (2018b). *Jumping into Civic Life: Stories of Public Work from Extension Professionals.* Dayton, OH: Kettering Foundation Press.

Peterson, W. (1938). Women's Place in Program Planning. *Extension Service Review*, 9(11), 161.

Phifer, B. M., List, E. F., & Faulkner, B. (1980). History of Community Development in America. In J. A. Christenson & J. W. Robinson (Eds.), *Community Development in America* (pp. 18–37). Ames: Iowa State University Press.

Price, M. (1938). The Fundamental Objectives of Agricultural Extension Work. Paper presented at the Supplement to the 1938 Agricultural Extension Conference. University of Minnesota. Department of Agriculture. Agricultural Extension Division, Minneapolis, MN.

Program Study and Discussion Section USDA. (1937). *Minutes [of the] Discussion-Group Training School for Negro District Agents.* State College, Petersburg, VA.

Reid, D. A. (2007). *Reaping a Greater Harvest: African Americans, the Extension Service, and Rural Reform in Jim Crow Texas.* College Station: Texas A&M University Press.

Rhoten, D. R., & Powell, W. W. (2011). Public Research Universities: From Land Grant to Federal Grant to Patent Grant Instutions. In D. R. Rhoten & C. Calhoun (Eds.), *Knowledge Matters: The Public Mission of the Research University* (pp. 315–341). New York: Columbia University Press.

Rudolph, F. (1962). *The American College and University, a History.* New York: Knopf.

Schön, D. A. (1995). Knowing-in-Action: The New Scholarship Requires a New Epistemology. *Change: The Magazine of Higher Learning*, 27(6), 27–34. doi:10.1080/00091383.1995.10544673.

Scott, R. V. (1970). *The Reluctant Farmer: The Rise of Agricultural Extension to 1914*. Urbana: University of Illinois Press.

Shaffer, T. J. (2012). The Land Grant System and Graduate Education: Reclaiming a Narrative of Engagement. In A. Gilvin, G. M. Roberts & C. Martin (Eds.), *Collaborative Futures: Critical Reflections on Publicly Active Graduate Education* (pp. 49–74). Syracuse, NY: Graduate School Press of Syracuse University.

———. (2017a). Democracy as Group Discussion and Collective Action: Facts, Values, and Strategies in Canadian and American Rural Landscapes. *The Good Society*, 26(2–3), 255–273.

———. (2017b). The Politics of Knowledge: Challenges and Opportunities for Social Justice Work in Higher Education Institutions. *eJournal of Public Affairs*, 6(1), 11–42. doi:10.21768/ejopa.v6i1.127.

———. (2017c). Supporting the "Archstone of Democracy": Cooperative Extension's Experiment with Deliberative Group Discussion. *Journal of Extension*, 55(5), 5FEA1. https://archives.joe.org/joe/2017october/a1.php.

———. (2018a). A Historical Note: Farmer Discussion Groups, Citizen-Centered Politics, and Cooperative Extension. In S. J. Peters, T. R. Alter & T. J. Shaffer (Eds.), *Jumping into Civic Life: Stories of Public Work from Extension Professionals* (pp. 191–212). Dayton, OH: Kettering Foundation Press.

———. Thinking Beyond Food and Fiber: Public Dialogue and Deliberation in the New Deal Department of Agriculture. In A. B. Hoflund, J. Jones & M. C. Pautz (Eds.), *The Intersection of Food and Public Health: Examining Current Challenges and Solutions in Policy and Politics* (pp. 307–326). New York: Routledge.

———. (2019a). Democracy in the Air: Radio as a Complement to Face-to-Face Discussion in the New Deal. *Journal of Radio & Audio Media*, 26(1), 21–34. doi:10.1080/19376529.2019.1564996.

———. (2019b). Enabling Civil Discourse: Creating Civic Space and Resources for Democratic Discussion. In R. G. Boatright, T. J. Shaffer, S. Sobieraj & D. G. Young (Eds.), *A Crisis of Civility? Political Discourse and Its Discontents* (pp. 188–209). New York: Routledge.

Sherwood, J. E. (2004). The Role of the Land-Grant Institution in the 21st Century. UC Berkeley: Center for Studies in Higher Education. http://escholarship.org/uc/item/1dp6w2cc.

Simon, L. A. K. (2010). Engaged Scholarship in Land-Grant and Research Universities. In H. E. Fitzgerald, C. Burack & S. D. Seifer (Eds.), *Handbook of Engaged Scholarship: Contemporary Landscapes, Future Directions* (Vol. 1: *Institutional Change*, pp. 99–118). East Lansing: Michigan State University Press.

Smith, C. B. (1938). On Turning the Page. *Extension Service Review*, 9(12), 177, 190.

Smith, C. B., & Wilson, M. C. (1930). *The Agricultural Extension System of the United States*. New York: John Wiley & Sons.

Smith, L. G. (1938). Teaching Floriculture and Ornamental Horticulture in New York State Through the Use of Local Leaders. Master's thesis, Cornell University.

Smith, R. G. (1949). *The People's Colleges: A History of the New York State Extension Service in Cornell University and the State, 1876–1948*. Ithaca, NY: Cornell University Press.

Spikes, D. R. (1992). Preface. In R. D. Christy & L. Williamson (Eds.), *A Century of Service: Land-Grant Colleges and Universities, 1890–1990* (pp. vii–ix). New Brunswick, NJ: Transaction Publishers.

Thomas, R. R. (2020). *Call My Name, Clemson: Documenting the Black Experience in an American University Community*. Iowa City: University of Iowa Press.

U.S. Extension Service. (1988). *Serving People in Need: Cooperative Extension at the 1890 Land-Grant Universities*. http://books.google.com/books?id=SJaNoopzzSsC.

Vest, C. M. (2007). *The American Research University from World War II to World Wide Web: Governments, the Private Sector, and the Emerging Meta-University*. Berkeley: University of California Press.

Von Engeln, O. D. (1924). *Concerning Cornell* (3rd ed.). Ithaca, NY: Cornell Co-Operative Society.

Wade, E. (2021). The History of HBCUs: Lessons on Innovation from the Past. In G. B. Crosby, K. A. White, M. A. Chanay & A. A. Hilton (Eds.), *Reimagining Historically Black Colleges and Universities: Survival Beyond 2021* (pp. 5–13). Bingley: Emerald Publishing.

Warburton, C. W. (1930). Six Million Farms as a School. *Journal of Adult Education*, 2(3), 289–293.

Weissman, S. (2021). A Debt Long Overdue. *Inside Higher Ed*, April 26. https://www.insidehighered.com/news/2021/04/26/tennessee-state-fights-chronic-underfunding.

Westbrook, J. R. (2010). Enhancing Limited-Resource Farmers' Economic, Environmental, and Social Outcomes Through Extension Education. PhD dissertation, Virginia Tech.

Wilson, M. L. (1935). Discussion Time Is Here. *Extension Service Review*, 6(10), 145.

Wood, S. D. (2006). The Roots of Black Power: Land, Civil Society, and the State in the Mississippi Delta, 1935–1968. PhD dissertation, University of Wisconsin–Madison.

Moving Beyond the Minimum Standard: Nondiscrimination Regulations, Policies, and Procedures

Norman E. Pruitt and Latoya M. Hicks

In this chapter, we will discuss U.S. nondiscrimination laws, regulations, policies, and procedures and how they interconnect with the land-grant's universities operations. In particular, our focus is to review the various civil rights laws and how they set a minimum standard for diversity, equity, and inclusion (DEI) for Extension operations. More importantly, given Extension's role in the community and representing the university's forward strides in social justice, we will examine how Extension can move beyond this minimum standard and move the needle on DEI to create equitable pathways for social justice and other forms of affirmative progressions for any and all groups of individuals who benefit from federal financial assistance.

The land-grant mission and understanding more specifically what constitutes a land-grant institution, the dynamics of its makeup, the recipients' expectations for creating "equal opportunity," and overlapping legal compliance can identify the premise associated with its operating constructs. The Morrill Acts of 1862 and 1890 defined the existence of the "land-grant" designation and the 1994 incorporation. More importantly, institutional funding sources are comprised of multiple facets of income such as contracts, grants, private donations, state resources, and county funding.

The acceptance of such financial assets creates a relationship and adherence to policies and/or practices. Federal laws, commonly referred to as nondiscrimination laws, set a minimum standard for civil rights; such laws include, but are not limited to, Title VI and Title VII of the Civil Rights Act of 1964, Title IX of the Education Amendments Act of 1972, sections 503, 504, and 508 (as applicable) of the Rehabilitation Act of 1973, and the Americans with Disabilities Act of 1990, as amended 2008. Additionally, institutions of higher learning are also subject to presidential executive orders such as 11246–Equal Employment Opportunity, 12250–Leadership and Coordination of Nondiscrimination Laws, and 13166–Improving Access to Services for Persons with Limited English Proficiency. For the purpose of this chapter, we focus on U.S. civil rights laws, including the Code of Federal Regulations (CFRs), presidential executive orders (EOs), and U.S. Department of Agriculture (USDA)/National Institute of Food and Agriculture (NIFA) policies and directives that directly apply to Extension programs and employment. Other U.S., state, and local laws, regulations, and policies may apply at the local program level.

Providing Educational Information to the People, from Civil Rights to Inclusion

Cooperative Extension is an outgrowth of the 1862 mission of the USDA to provide practical research to the people. The idea of providing access to a wealth of information produced by land-grants and to educate the people has moved us from segregation toward the ultimate goal, some might say, of full inclusion and equity in Extension programs to include *all* people on a nondiscriminatory basis. Today, NIFA is tasked with effectuating the USDA's mission of teaching, research, and Extension by partnering primarily with state land-grant universities.

Direct (intentional) discrimination by federal, state, or local governments is prohibited by the Constitution; however, indirect (unintentional) discrimination using federal funds is just as invidious and covered by applicable laws. President John F. Kennedy asserted the concept of equal access in federal assisted programs, such as land-grants, as "simple justice" when he presented the Civil Rights Act and his message to Congress on June 19, 1963: "Simple justice requires that public funds, to which all taxpayers of all races [colors, and national origin] contribute, not be spent in any fashion which encourages, entrenches, subsidizes or results in racial [color or national origin] discrimination" (DOJ, 2019b, para. 2). However, justice has never been simple—hence the enactment of federal civil rights laws to act as a combative aid and advocate for basic human liberties. Understanding the foundational principles that historic U.S. civil rights laws represent creates birthed opportunities for expansions of such laws to further ensure the necessary inclusion of individuals.

The Role of Extension in Promoting Equitable Policies Locally, Statewide, and Nationally

Land-grant institutions are tasked with providing informal and practical research-based information to potential and eligible individuals within the United States (and those locations defined as Insular Areas). The role of providing education to the people of the United States places Extension as the gatekeeper with regard to efforts at equality, social justice, and removing potential barriers of discrimination. The Extension agent, educator, specialist, paraprofessional, and other university and institutional faculty and staff are the backbone of disseminating such information to individuals and tasked with delivering educational programs funded by governmental partnership on a nondiscriminatory basis. The multifaceted relationship is thus characterized as a three-legged stool of teaching, research, and Extension providing access to all potential and eligible audiences.

The social justice and civil rights events of 2020 and 2021 identified additional systematic and structural inequities across the United States and have allowed many organizations, including universities, an opportunity for reaffirmed commitments to diversity, equity, and inclusion efforts. Given Extension's multifocal operation to include such presence at the local level, fostering such community relationships as an expressed vision of its universities' renewed focus on social justice, inequities, and civil rights is paramount for an Extension entity. Many communities across the United States are expressing various forms of disparities from relationship building, trust, and empathy to a lack of educational empowerment (Lempinen, 2021; EdSource, 2021). Reexamining equitable policies and practices as a form of preventative planning provides Extension this linkage within the communities it services, and provides its clientele the necessary objectives of sponsorship, allyship, and advocacy and thus accessibility to statewide nonformal education.

Extension has emerged as a critical component of delivering a myriad of educational programs among community benefactors. According to the U.S. Government Accountability Office, "The Cooperative Extension Service is the largest education system of its kind in the world . . . established in 1914 primarily to provide farmers with information from agricultural research" (GAO, 1981, para. 1). Extension's evolution as a subset entity within a college of agriculture provides a vast level of learning, community engagement, civic dialogue, and training to broad audiences within a particular region. Sometimes these activities are referred to as extramural studies, adult education, youth development, and other forms of information sharing. As the importance of higher education continues to grow, particularly in suburban and rural communities, significantly it has become more difficult for individuals to participate in a concerted effort and collectively acquire invaluable educational programming designed to enhance quality living (Leonard, 2012). Operating under the umbrella of their institutions, Extension operations are required to follow at a minimum the differentiating laws that apply as recipients of federal financial assistance and charged with the corporate responsibility to develop, implement, and deliver programming in a nondiscriminatory manner.

The significance of the USDA/NIFA's role in supporting agriculture in higher education and the accompanying civil rights responsibility at a land-grant institution cannot be understated. Public Law 107-293 establishes "firmly the Department of Agriculture as the lead agency in the Federal Government for the food and agricultural sciences, and [emphasizes] that agricultural research, extension, and teaching are distinct missions of the Department of Agriculture" (NARETPA, 1977, §3102). The unique funding and historical role of Extension within the USDA was to provide agricultural information to all eligible and potential audiences; hence, a specific responsibility for Extension staff is assuring and making positive efforts (often identified as affirmative action or all reasonable efforts) to deliver programs to a wide range of clientele.

Federal Nondiscrimination Laws

There is a vast number of nondiscrimination laws affecting the day-to-day operation of the Extension director or 1890 administrator, educator/agent, specialist/university faculty member, staff, and in many cases, volunteers in such programs as 4-H and Master Gardeners. In this section we review the federal laws (acts) that have the most direct effect on Extension programs and employment (figure 1). The federal nondiscrimination laws are the minimum standard for Extension programs; however, our goal as leaders is to go beyond the minimum nondiscrimination standards where applicable. Other federal laws that may affect Extension programming are the Age Discrimination Act of 1975 (Section 6101), the Equal Pay Act of 1963, and the Food Stamp Act of 1977 (Supplemental Nutrition Assistance Program Education [SNAP-Ed]).

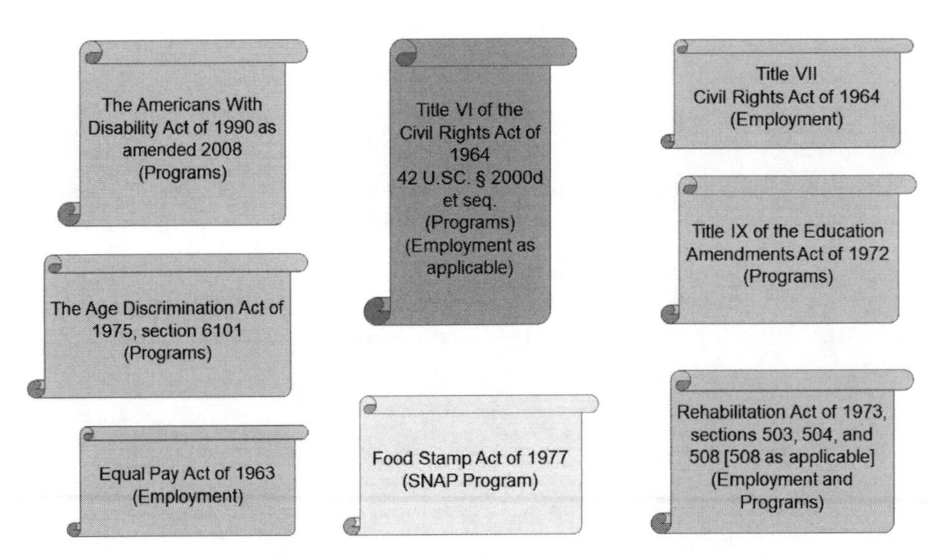

FIGURE 1. Overview of Nondiscrimination Laws. The governing authorities and the point of intersection in Extension programs and employment. *Norman E. Pruitt and Latoya M. Hicks.*

There are a number of states at the local and university levels adopting common laws, regulations, policies, and processes that meet those identified within federal regulations; however, some states have exceeded the minimum federal civil rights nondiscrimination standards. Universities and states routinely have extended protections for nondiscrimination above federal laws, such as safeguards afforded to lesbian, gay, bisexual, transgender, and queer (LGBTQ) communities (other forms of gender identity and expression, etc.), that are included in most universities'/states' nondiscrimination policies and statements. For example, the University of Maryland nondiscrimination policy notes:

> This Policy prohibits discrimination on grounds protected under Federal and Maryland law and Board of Regents policies. University programs, activities, and facilities are available to all without regard to race, color, sex, gender identity or expression, sexual orientation, marital status, age, national origin, political affiliation, physical or mental disability, religion, protected veteran status, genetic information, personal appearance, or any other legally protected class. (UMD, 2019, para. 1)

The University of Maryland nondiscrimination policy includes protections under Maryland state law and university board of regents policies that are signed by the institution's president and effectuated by executives, faculty, and staff members. Extension operations are also required to provide notification and inform clientele of their rights under federal law including the right to file a complaint with the secretary of agriculture. For example, the "And Justice for All" posters serve as notification of the complaint procedure for the USDA and are to be posted and utilized by Extension staff (NIFA, 2019).

The notification process also takes other forms such as nondiscrimination statements that are typically on Extension publications, notifications of free language and disability accommodations, and assurance statements from organizations or groups that Extension partners with to deliver educational content. Extension operations must also ensure that educational training and awareness is provided to its staff on a continual basis and understand that not all protections are extended to its clientele under federal law relative to institutional policies and/or state/local laws and provisions. Therefore, it is necessary to develop broadening relationships with institutional university level offices, state attorney general's offices, and other compliance platforms as needed for specific issues related to program and civil rights awareness and to provide programing beyond the minimum nondiscrimination standards of the federal government.

Why civil rights laws? Why do we need civil rights laws? Why are they important and who do they really protect? The necessity of such laws can be found in assuring there is at least a minimum level of support, redress, and protection at the federal level for equal treatment of our Extension clientele and employees. The noted basic human rights have allowed individuals to provide some limited protections for themselves, families, and businesses. Without the enactment of such laws, how can we truly measure equality or accountability within the behavior of people or systems despite its design and/or distinctions? Utilizing 2020 and 2021 as a point of reference, the only remedy to social injustice or discrimination is through legal action, federal review, and/or consent decree. Hence, the most important tool to fight injustice and fight for equal access when all other constructs, afforded provisions, or state and local laws fail are the civil rights laws of these United States.

Title VI of the Civil Rights Act of 1964—Program Nondiscrimination

Public Law 88-352 (H.R. 7152), hereinafter the Civil Rights Act of 1964, is the overarching law covering nondiscrimination in the United States that applies to institutions receiving federal financial assistance such as Extension land-grant programs. The act has eleven titles, with Title VI codified at 42 U.SC. § 2000d et seq (Civil Rights Act, 1964). However, Title VI, which addresses civil rights for recipients of federal financial assistance, was one of the most discussed assertions. Title VI of the Civil Rights Act of 1964 and the acceptance of federal funds require Extension organizations, Extension executives and personnel, Extension volunteers, and subrecipients of Extension to adhere to the provisions and legal guidelines of nondiscrimination in the delivery of educational programming.

Title VI defines the term "program or activity" as "all of the operations of a college, university, or other postsecondary institution, or a public system of higher education" (DOJ, 2020, p. 27). The Civil Rights Restoration Act of 1987 clarifies what constitutes a program or activity, and as defined by such act, all of the operations in receipt of federal funding come under the sphere of federal nondiscrimination laws (Civil Rights Restoration Act, 1987), and such scrutiny/compliance may be determined by the funding agency (USDA/NIFA). In our travels around the United States, Extension employees often explained "our program is county funded" as if to say the civil rights laws do not apply. It is extremely important to note that the Extension director or 1890 administrator, as a recipient of federal funds (regardless of the funding streams [i.e., state match, county contributions, donations, etc.]), is ultimately responsible for civil rights, and thus its employees are indirect recipients—therefore such responsibility is shared throughout the organization.

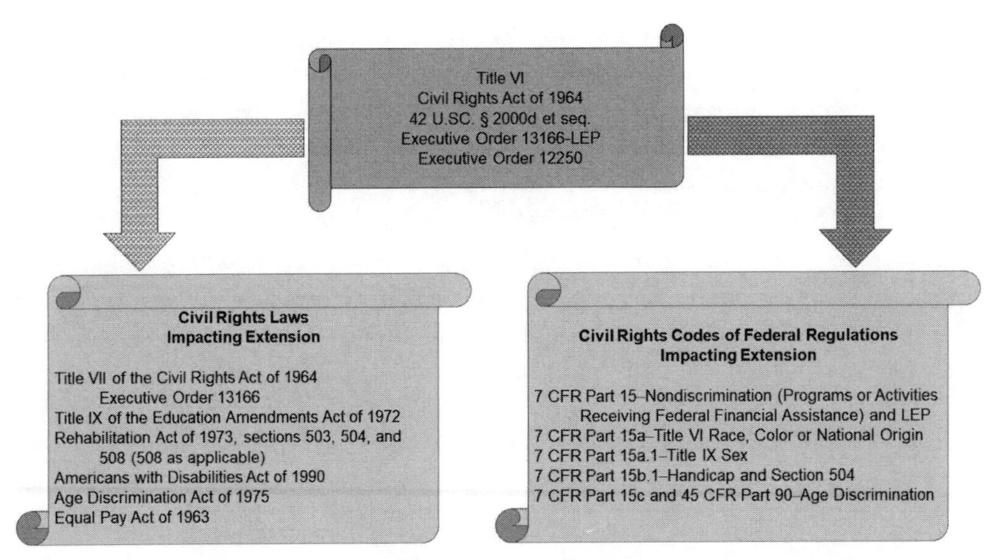

FIGURE 2. Title VI of the Civil Rights Act Program Provisions. The regulatory guidance effecting programs and services delivered on behalf of Extension. *Norman E. Pruitt and Latoya M. Hicks.*

While Title VI of the act prohibits discrimination on the basis of race, color, and national origin (Civil Rights Act, 1964, p. 252), the act also allows for individual agencies such as USDA/NIFA with the Department of Justice (DOJ) to effectuate the Civil Rights Act of 1964 and issue agency guidance, regulations, policies, and procedures through a utilization of Codes of Federal Regulations (CFR). Directives of the USDA that govern the operating framework of federally assisted programs begin with the number 7 (figure 2). The USDA therefore issued 7 CFR 15 subpart A, "Nondiscrimination in Federally-Assisted Programs of the Department of Agriculture-Effectuation of Title VI of the Civil Rights Act of 1964."

The USDA further provides departmental regulations (DR) such as 4330-002 to support Title VI law and 7 CFR 15. USDA 7 CFR 15 and part 16 has several key aspects highlighting the responsibilities of the Extension director/1890 administrator and Extension personnel delivering programs (LII, 2019), such as:

- *Program discrimination prohibited*—7 CFR expressly prohibits discrimination in (1) "making available instructions, demonstrations, information, and publications"; (2) "the use in any program or activity funded by the Cooperative Extension Service of any facility, including offices, training facilities"; and (3) access to training, participation in fairs, competitions, field days, etc.
- *Assurance*—Extension must provide assurance of compliance with the laws and nondiscrimination in programs and employment. Hence, the collection of program and employment data is crucial.
- *Compliance*—Extension is subject to USDA/NIFA civil rights compliance reviews. Extension must maintain records to assure compliance.

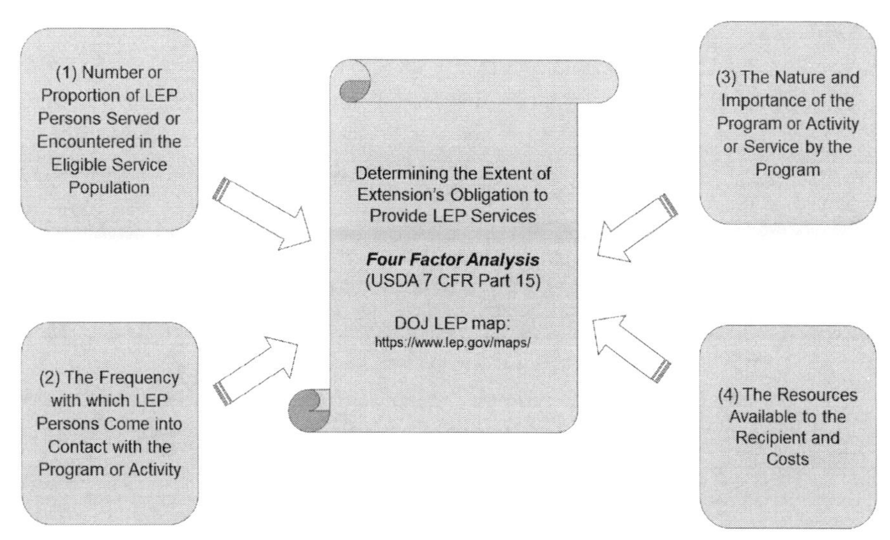

FIGURE 3. Four Factor Analysis. Identifying LEP personnel through an interpretative method.
Norman E. Pruitt and Latoya M. Hicks.

- *Employment*—Extension employment practices are subject to 7 CFR 15 where individuals are not excluded from participating in employment activities or denied related employment benefits on the basis of race, color, sex, or national origin as it relates to the delivery of programs.
- *Religion*—An organization that participates in programs and activities supported by direct USDA assistance programs shall not discriminate against a program beneficiary or prospective program beneficiary on the basis of religion or religious belief.

In addition to understanding how Title VI extends to the development of DRs/CFRs, it is also important to recognize the role of U.S. presidents when issuing EOs to further define actions to be taken into account with civil rights laws, to clarify the intent of civil rights laws, and/or to extend coverage of said laws. President Joseph Biden on January 20, 2021, issued EO 13985-Executive Order on Advancing Racial Equity and Support for Underserved Communities through the Federal Government, which supports current diversity, equity, and inclusion efforts at universities and Extension operations. Another EO that had an effect on Extension programming and the linkage of the law to Extension programs was President William J. Clinton's EO 13166 (April 16, 2000), which requires "Improving Access to Services for Persons with Limited English Proficiency" (LEP). The overall requirements for language access in Extension programs are established under national origin discrimination—Title VI of the Civil Rights Act of 1964 and EO 13166.

The DOJ and subsequently the USDA issued LEP guidance via 7 CFR 15, and NIFA issued LEP guidance to land-grants and other institutional partners on April 15, 2016. Extension services must conduct an LEP self-assessment and develop an LEP implementation plan utilizing the "Four LEP Factors" (DOJ, 2019a) to determine into which languages to translate materials and provide interpretation services (figure 3).

There are additional language requirements as well as civil rights requirements enforced at the state and local levels for Extension operations such as the provision within the Supplemental Nutrition Assistance Program Education. Specifically, when addressing the civil rights compliance of SNAP-Ed, the Food Stamp Act of 1977 identifies specific program guidelines, and these are extended to issues not covered under Title VI, thus including sex, religious creed, or political beliefs (Food Stamp Act, 1977). Utilizing the state of Maryland as a point of reference, a SNAP-Ed program would also comply with the provisions of the Maryland Equal Access to Public Services Act of 2002, which requires the translation of vital information to populations of 3.00 percent within a geographical location.

Title VII of the Civil Rights Act of 1964—Employment Nondiscrimination

The legal framework enabling organizations to take affirmative employment approaches is Title VII of the Civil Rights Act of 1964. Organizations are therefore prohibited from participating in unlawful employment practice by failing or refusing to hire or discharge any individual, or otherwise to discriminate against any individual with respect to his/her compensation terms, conditions, or privileges of employment because of such individual's race, color, religion, sex, or national origin (Civil Rights Act, 1964, p. 255). Title VII's overarching authority is extended to Extension if such operating guidelines attempt to limit, segregate, or classify employment in any way that would deprive or tend to deprive an individual of employment opportunities or otherwise adversely affect his/her status as an employee on such bases noted earlier.

EO 11246 and related employment laws further expand employment expectations for universities (Extension), operating on the premise of receiving federal funding and being identified as federal contractors. The U.S. Department of Labor manages such civil rights efforts and notes, "The contractor will not discriminate against any employee or applicant for employment because of race, color, religion, sex, sexual orientation, gender identity, or national origin. The contractor will take affirmative action to ensure that applicants are employed, and that employees are treated during employment, without regard to their race, color, religion, sex, sexual orientation, gender identity, or national origin" (DOL, 2019, sec. 202.1).

EO 11246 provides extended protections in employment for sexual orientation and gender identity that was not specifically implied within the Civil Rights Act of 1964 until recent Supreme Court ruling on protections June 15, 2020—Bostock v. Clayton County. Therefore, the federal government may increase protection against employment discrimination by issuing executive orders or other relative regulations to expound upon narrowly focused guidance affecting a particular audience. As previously noted, organizations and institutions occasionally adopt policies and practices "examining organizational culture, culture differences, cultural change and cross-cultural relationships" (Ewoh, 2013, p. 113) to spearhead inclusive objectives and promote leading change.

The USDA aligns Extension employment activities to the delivery of programs; given the overwhelming expenditure of Extension budgets on employees in the form of compensation and benefits, the inability to perform employment-related activities coupled with discriminatory practices would potentially adversely impact the effectiveness of an Extension operation. USDA 7 CFR 15 notes:

> Where a primary objective of the Federal financial assistance is not to provide employment, but discrimination on the grounds of race, color, or national origin in the employment practices of the recipient or other persons . . . to exclude individuals from participation in, to deny them the benefits of, or to subject them to discrimination under any program or activity . . . the foregoing provisions of this §15.3(c) shall apply to the employment practices of the recipient or other persons. (LII, 2019, §15.3[c])

Title IX of the Education Amendments Act of 1972—Nondiscrimination on the Basis of Sex

Nondiscrimination in Extension programs based on sex is the hallmark of Title IX of the Education Amendments Act of 1972. The act explicitly identifies higher education, and hence Extension performed at declared institutions are required to adhere to the act. The U.S. Department of Education usually takes the principal role enforcing Title IX programs. The USDA expressly identified the same protections for individuals in Extension programs with the issuance of 7 CFR 15a, subpart d: "Except as provided elsewhere in this part, no person shall, on the basis of sex, be excluded from participation in, be denied the benefits of, or be subjected to discrimination under any academic, extracurricular, research, occupational training, or other education program or activity operated by a recipient that receives Federal financial assistance" (LII, 2019b). Extension 4-H programs more specifically require special attention due to the nature and size of such programming, the statewide presence, and the reliance on volunteers to carry out its effective delivery. Extension operations must also take a proactive notion to ensure federally assisted programs are accessible to the public and/or the same level of equal access is provided.

The Rehabilitation Act of 1973 (as Amended)—Disability Access

The Rehabilitation Act of 1973 (Public Law 93-112) identifies several sections within the law, such as sections 503, 504, and 508, that provide individuals with disabilities access to programs and employment. Section 503 of the Rehabilitation Act of 1973 provides access to employment and advancement for disabled individuals employed by a federal contractor (university) with a contract in excess of $10,000. Section 504 notes, that "no otherwise qualified handicapped individual in the United States . . . shall . . . be excluded from the participation in, be denied the benefits of, or be subjected to discrimination under any program or activity receiving Federal financial assistance" (Rehabilitation Act, 1973, p. 394). There are monetary thresholds and technical focuses separating the two sections: employment and structural access for Extension operations. However, both sections apply to Extension enterprises as an employer and program provider. Further considerations for Extension organizations are to understand how both sections can intersect at a given time. For example, Extension agents/educators who are hired and housed in nonuniversity facilities, which is typical of county offices, need to consider accommodations for employees with disabilities and can extend to programs with regard to delivering educational content and/or being a beneficiary of an educational program.

Institutions are not subjected to restrictions under section 508 under the Rehabilitation Act of 1973 as it is a federal agency requirement per se (Rehabilitation Act, 1973b). However, ensuring technical standards are in place to provide a greater level of access and opportunity to recipients participating in federal programming creates a pathway to access. More importantly, institutions are subjected to the conditions set forth in the Americans with Disabilities Act of 1990, as amended in 2008 (Americans with Disabilities Act Amendments Act (ADAAA), which requires such entities to ensure a method of communication regardless of the methodology or content and develop a road map of equitable opportunity. The ADA is the nation's first comprehensive civil rights law addressing the needs of individuals with disabilities, prohibiting discrimination in employment, public services, public accommodations, and telecommunications (Americans with Disabilities Act, 1990, para. 1).

Section 508 of the Rehabilitation Act was enacted to eliminate barriers imposed on individuals with disabilities (IWD) and the general public regarding the accessibility to electronic information technology. Such enactment streamlined an opportunity for IWD and the general public to acquire information from web-based material or within a facility in the same comparative manner available to other individuals. Extension operations should provide notice to employees and applicants with disabilities of their rights to access on technological platforms but more importantly work to remove barriers that may exist to the general public, employees, and applicants.

Obligation of Universities to Reach Underserved, Underrepresented, and Socially Disadvantaged Populations

Land-grant universities, from their inception in 1862 and 1890, were charged with the mission to provide education to the people of these United States—hence, the College of Agriculture's referral to as the "people's college." The acceptance of federal funding by states and the creation of universities to carry out the vision of President Abraham Lincoln, Senator Justin Morrill, and the USDA created the legal and strategic obligation and mission to service the people of these United States and respective territories. The National Agricultural Research, Extension, and Teaching Policy Act of 1977 further promoted the obligation of land-grants to service a broad range of diverse audiences, and the purpose of federally supported agricultural research, extension, and education is to:

> (1) enhance the competitiveness of the United States agriculture and food industry in an increasingly competitive world environment, (2) increase the long-term productivity of the United States agriculture and food industry while maintaining and enhancing the natural resource base on which rural America and the United States agricultural economy depend, (3) develop new uses and new products for agricultural commodities, such as alternative fuels, and develop new crops, (4) support agricultural research and extension to promote economic opportunity in rural communities and to meet the increasing demand for information and technology transfer throughout the United States agriculture industry, (5) improve risk management in the United States agriculture industry, (6) improve the safe production and processing of, and adding of value to, United States food and fiber resources using methods that maintain the balance between yield and environmental soundness, (7) support higher education in agriculture to give the next generation of Americans the knowledge, technology, and applications necessary to enhance the competitiveness of United States agriculture, (8) maintain

> an adequate, nutritious, and safe supply of food to meet human nutritional needs and requirements, and (9) support international collaboration that leverages resources and advances priority food and agricultural interests of the United States. (NARETPA, 1977)

Extension services are the heartbeat of a national system that supports the United States' interest to provide food to all individuals and to also provide a transfer of research information on a nondiscriminatory basis. This obligation also extends to those individuals who may be identified as underserved, underrepresented, and/or socially disadvantaged.[1] Individuals or groups identified as underserved, underrepresented, and/or socially disadvantaged have no specific race, color, national origin, ethnicity, gender, etc. The USDA has also included incentives and programs to improve access for such audiences typically not participating in sponsored programs at such noticeable rates. For example, the 2008 Farm Act includes participation incentives for beginning limited-resource and socially disadvantaged farmers and ranchers regardless of race, ethnicity, gender, etc.

The U.S. population continues to change over time, and Extension operations must adapt to such change and mirror programs that are attractive to beneficiaries. The diversity of the population in 1914 at the beginning of Smith-Lever funds is far different than the population noted in the 2020 census. When the USDA and the land-grant university system were founded in 1862, the 1860 U.S. decennial census counted 31,443,321 people in the United States. The dominant racial categories were White (85.6 percent) and Black (14.1 percent) (Gibson & Jung, 2002, No. 56, Table 1). The 2020 census shows the United States has grown to 331,449,281 of which there are seven racial categories and Hispanics are counted as an ethnicity (U.S. Census, 2020a).

Populations are also grouped in USDA and Extension programs as underserved, underrepresented, and/or socially disadvantaged. The USDA has defined these terms in various legislation such as the Farm Bill or in other government programs such as those targeted to small business. Underserved (SUTA, 2012; OMB, 2014), underrepresented (NIFA, 2017), and/or socially disadvantaged (NSAC, 2019; FACT, 1990, pp. 128–129) were initially focused on race and ethnicity; gender was added to socially disadvantaged populations with Public Law 102-554 (CFRDA, 1992). For the purpose of our conversation in this chapter the definitions for underserved, underrepresented, and socially disadvantaged are included in the note.

Potential, Eligible, and Actual Extension Program Clientele

Our discussions relative to identifying potential and eligible clientele will focus on the collective of all individuals including those identified as underserved, underrepresented, and/or socially disadvantaged whose race, ethnicity, and/or gender or other federally protected status has perpetuated or created their current access or lack of access to Extension programs. All programming starts with the concept of identifying a potential audience for programming. Extension, like any institution or business as part of its operations, should be familiar with its audience and in particular key demographics of potential audiences. In our travels as compliance officials, it was not uncommon for Extension leaders, agents/educators, and/or staff to not be familiar with either the demographics of their audiences or, more importantly, where to find such information.

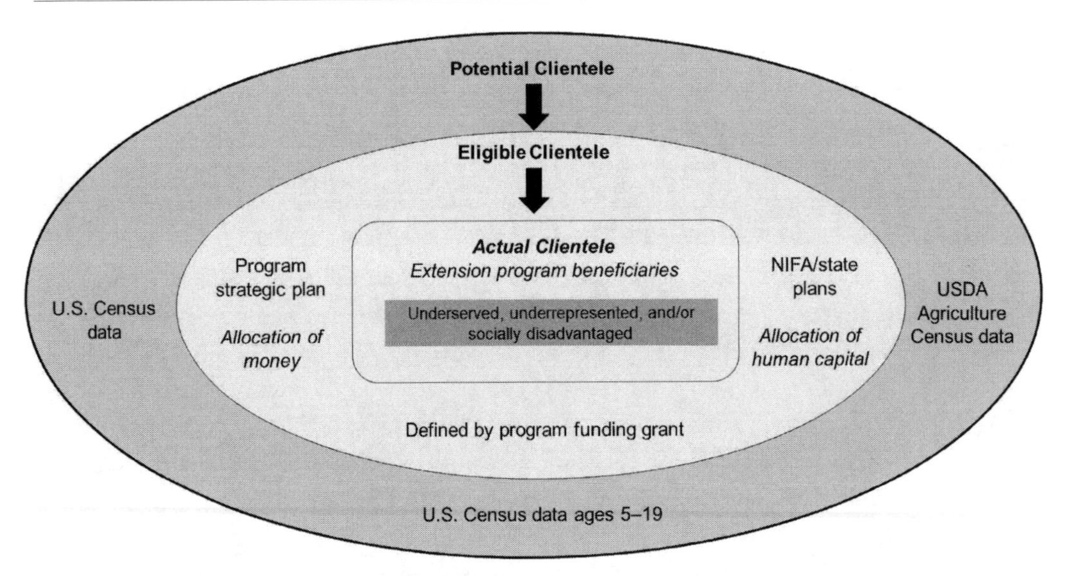

FIGURE 4. Variations of Programmatic Audiences. Extension populations as defined by programs and outreach criteria. *Norman E. Pruitt and Latoya M. Hicks.*

The key official U.S. government demographics benchmarks we utilize for this chapter are the U.S. decennial census (DP-1 and P-files), the Census of Agriculture, and the U.S. Department of Justice Limited English Proficiency data.

We begin the concept of identifying audiences for the purpose of programming and obtaining funding to support the overall potential clientele. The potential as shown in figure 4 captures the broadening concept of all audiences despite eligibility characteristics. For Extension to determine those eligible, several factors can isolate such variables. There are four key organizational or funding factors for land-grant Extension programs noted in figure 4 that typically determine the totality of the eligible clientele base: Extension program specific strategic plan, current resources or capacity, NIFA/State Plan of Work, and program funding grant agreements. For the purpose of this discussion, let's determine the broad audience and programmatic areas of Maryland.

In the 2020 Maryland U.S. Census, the total potential number of clientele was 6,177,224; in the case of agricultural programs, the 2017 Maryland Census of Agriculture (Table 63—producer characteristics with maximum of 4 producers by farm) notes there were 21,279 producers (formerly operators) including 16,879 primary producers (formerly principal operators) per farm (USDA, 2017). There are some agricultural and other program areas with more than one potential audience, such as the Maryland producers (21,279 if educational programs for producers or 16,879 if the program focus is the primary producer) who by definition are farmers and ranchers. The eligible population (figure 4) may be the same as the potential, but typically it is a subset and usually based on the program's mission.

However, the potential audience for residential horticulture programs, such as the clientele served by Master Gardeners, would be the general population for the state of Maryland (6,177,224). The population of clientele eligibility may also depend on specific program emphasis such as operators focused on agronomic crops, commercial horticulture, livestock, organic, small farmer, etc.

The eligible population for advisory groups may also be determined by program emphasis, such as primary producers being utilized for agriculture advisory groups, setting policy, or program direction.

Grant programs that are operational in many 1862 and 1890 Extension programs such as the Expanded Food and Nutrition Education Program (EFNEP) and SNAP-Ed have a program focus of nutrition, narrowing program eligibility to certain economic populations and/or grant specific for these educational programs. Other specific USDA grant programs determining eligibility include, but are not limited to, the NIFA Beginning Farmer and Rancher development program and the Organic Agriculture Research and Extension Initiative, which may limit eligible audiences to geographic boundaries or socioeconomic status. However, such requirements and selections of program populations to be served must be done on a nondiscriminatory basis.

The *actual* beneficiaries of Extension programs are those individuals who directly receive educational benefits from Extension. The USDA defines the beneficiary as "a person or group of persons with an entitlement to receive or utilize the benefits, services, resources, and information, or participate in activities and programs conducted or funded in whole or in part by USDA" (USDA, 1999). Notice the emphasis on "with an entitlement to receive . . . benefits." The beneficiary may utilize the benefits, resources, and information or participate in activities funded in whole or in part by the USDA. Hence, this covers land-grants that receive federal financial assistance where in most instances there is a state matching (financial funding) requirement.

Extension organizations must maintain a data-collection system to measure on both a quantitative and qualitative basis how populations are identified, how participants were notified and participated in programs, benefits, and received services, and how such program activities were delivered in a nondiscriminatory manner to both populations determined as minorities and nonminorities. The comparison of race, ethnicity, and gender of individuals receiving Extension benefits and services as compared to race, ethnicity, and gender of eligible individuals is a terminology known as "parity" for civil rights purposes. Parity is defined by NIFA's predecessor agency (USDA Extension Service), and "participation refers to a condition in which the percent distribution of program participants (beneficiaries) is proportionate to, or within reasonable limits of (± 2.0 percentage points) of their respective percent distribution in the potential recipient audience/population" (USDA Extension Service, 1983, p. 5). Extension operations should have the parity percentages identified in their respective affirmative action plans for Extension programs.

The concept of parity forms the basis of determining nondiscrimination in Extension programs and is more a legal concept that may or may not be a true measurement of inclusion, equality of services, and/or equal access. Other methods of analyzing potential, eligible, and actual data beyond the legal parity construct for inclusion and equity purposes should be utilized to diversify programs. For example, Extension operations may review the proportion of those actually reached by each race, ethnicity, and gender to those eligible by the same theory to determine equitable distribution of benefits and services. This comparison may be suitable for identifying and or uncovering biases (unconscious or conscious) in program delivery as well as future program outreach objectives and identifying various needs assessments for training. Regardless of the methods utilized and/or adopted by Extension operations to determine diversity in programs, the minimum federal legal standards must be adhered to and established assurances to equal access.

Methods for an Affirmative Approach to Crafting an Inclusive Workforce

Creating a diverse and inclusive workforce has extended beyond a legal requirement and has become an opportunity for organizations to dominate industries with such competitive advantage. Extension operations, in adopting and promoting such inclusivity by replicating the communities in which they serve, gain more accessibility to innovative ideas by employing diverse talent pools and strategically branding its organization as a champion of equality. Extension enterprises are unique, with the intersectional components of its organization operating under the umbrella of a land-grant institution and the structural design of its staff to provide programs and services to their benefactors. For instance, agents, specialists, paraprofessionals, and other faculty who specialize in particular backgrounds such as agriculture, horticulture, agronomy, sustainable food, and nutrition are some of the relevant requisite skills for Extension labor markets, and such talent is considered distinctive and/or scarce. Understanding the nuances of attracting and retaining unique abilities are barriers within itself to strategically plan for. Therefore, Extension directors/1890 administrators must create isolated opportunities to grow internally their own talent pools by building a model for succession planning to establish and forecast the skill bank needed for tomorrow and to construct a mechanism to diversify future staff and talent needs.

Affirmative action is an approach utilized for promoting equitable access for qualified individuals to acquire opportunities in any and all aspects of employment. Affirmative action is critical for addressing historical challenges as well as encouraging diversity within institutions; however, such strides to ensure access should not be misconstrued with withholding an opportunity from one group and redistributing it to another as a form of equity. In this regard, discrimination within employment activities can occur to any individual regardless of their protected status, and therefore it's imperative for Extension operations to conduct barrier analyses to determine if organizational practices, policies, and/or procedures have intentional or unintentional effects on their workforce.

Extension operations in general deliver programs and services to diverse communities across the state, and the effectiveness of such delivery of their benefits is determined by multiple facets such as establishing trust within communities, recruiting stakeholders who can advise Extension staff of programmatic and community needs, and learning the culture and norms of any environment (professional and/or personal) that would aid program sufficiency and research dissemination. The ideal diverse workforce within Extension operations includes men and women from various ethnic backgrounds, diverse age cohorts (Generation X and Generation Y), and individuals with disabilities (Bezrukova, Jehn & Spell, 2012). Extension units within a department (college) are subjected to the legal provisions set forth by DOL, more specifically the previously mentioned EO 11246, and must establish affirmative action goals as a federal contractor and recipient of federal funds.

Fully implementing an affirmative action plan and evaluating Extension's workforce analysis paints an illustrative overview of the organization and the demographics of its staff to include compensatory and occupational support. Hence, such information can determine where efforts to attract or retain diversity exist. Identifying action-oriented goals and the time line to complete said goals is instrumental and would take place when reviewing the job group analysis of Extension's affirmative action plan to determine the availability of a particular group's accession to leadership, to establish feeder groups for promotional opportunities as a result of job enrichment, enhancement, sharing, or rotational assignments. Additionally, Extension must isolate numerical expressions to establish a reasonable goal when comparing incumbency to the estimated availability of Extension employment activities. Universities should be prepared to take bold steps to increase diversity and prepare units how to be inclusive such as the Faculty Advancement at Maryland for Inclusive Learning and Excellence (FAMILE: UMD, 2021), a new diversity program launched in 2021 by the president of the University of Maryland, College Park.

Diversity, Equity, and Inclusion as the Engine of Opportunity

Diversity, equity, and inclusion initiatives within organizations emphasize incorporating qualified individuals from diverse backgrounds. Diversity alone exemplifies an organization's ability to express variation, and Extension organizations are challenged with implementing vigorous strategic plans to promote parity programmatically from both Title VI and VII fundamentals. "Managing diversity cannot simply be about business appeal" (Jones, 2006, p. 151); it must be measured and implemented subsequently with inclusivity in order to determine its effectiveness. A method to utilize diversity and inclusion as a path forward for Extension operations would include the following elements:

- establish benchmarks to revisit strategic initiatives, reassess structural barriers, and examine opportunities for access to employment or programs;
- integrate validity studies to monitor program effectiveness;
- analyze efforts to incorporate stakeholders' input and involvement;
- survey what resources are needed and sufficient for program delivery or employment attainment;
- train and evaluate outreach goals to determine a transfer of knowledge and if yielded results were achieved; and
- communicate culture transformations of inclusiveness and belonging.

Extension operations should not only consider utilizing diversity strategies but couple such approaches with inclusion efforts. An example of a DEI plan that encompasses the six items for a path forward is the University of Maryland, College of Agriculture and Natural Resources "Diversity, Equity, Inclusion, and Respect Plan" (2021; note that respect was added to the traditional concept of DEI). Diversity, equity, and inclusion issues have dominated headlines for many decades in the United States as the quest for equal representation continues to persist. While legal provisions such as Title VI of the Civil Rights Act of 1964 and others provide a roadmap toward the creation of an inclusive work environment, some organizations are yet to comply due to a combination of challenges. To fully move forward with diversity, equity, and/or inclusion efforts, Extension must first understand the demographics of the population. Affirmative strides to exceed the minimum federal standards on diversity require measuring demographic changes in program participation and/or the workforce to champion equality. Diversity and inclusion efforts are considered change elements for Extension operations when visiting the demographics of the United States.

The 2020 Census noted a U.S. population of 331,449,281 with the ability to collect individual surveyed data on seven racial categories, ethnicity, gender, age, language, disability, and veteran status, to name a few. Our diversity in this nation is substantial as individuals who identify their race as other than White make up 38.4 percent of the population 8.4 percent who identify themselves as "Some Other Race" and 10.2 percent identifying as "Two or More Races." Hispanics of any race make up 18.7 percent, and women account for 50.52 percent of the population (U.S. Census, 2020). However, does this translate into equal access to programs or employment, or would we consider women to be underserved and underrepresented in selected Extension programs? Questions frequently asked by Extension staff during our visits to Extension operations as NIFA compliance officers were the origin of the racial categories and whether Hispanic counted as a race. Racial categories have evolved over time within the United States, and such categories are defined by the U.S. Office of Management and Budget (U.S. Census, 2019).

Demographic data and studying such are not new methodologies in the United States. The interest in the racial and gender composition of the United States can be traced to colonial times. Prior to the first U.S. Census in 1790, estimates dating back to 1610 by individuals such as George Bancroft, an American historian, and Franklin Bowditch Dexter of Yale University as well as data estimates from the U.S. Census illustrate the growth of the American colonies. Bancroft estimated the colonies grew from 1,200,000 people of whom 1,040,000 were White and 220,000 were Black in 1750 to 2,945,000 people in 1780 of whom 2,383,000 were White and 562,000 were Black (Census, 2020b). The first census in 1790 noted the total U.S. population was 3,929,214 of which 3,172,006 (80.7 percent) were White and 757,208 (19.3 percent) were Black, and of the latter 697,681 (92.1 percent) were slaves (Gibson & Jung, 2002, Table 1). The 1860 census was the first census to collect data on races other than White and Black, which ascertained that of the 31,443,321 people in the United States, 85.6 percent were White, 14.1 percent Black, 0.1 percent American Indian, Eskimo, and Aleut, and 0.1 percent Asian and Pacific Islander. Hispanics were first reported in the 1940 census and made up 1.4 percent of the population; in 1950, the U.S. Census recognized "Other race" as a category (Gibson & Jung, 2002, Table 1).

The civil rights laws of the United States have evolved well beyond laws in 1875 to include extended protections. As we noted in this chapter, the initial laws focused on race and have thus evolved into multiple laws, regulations, and EO protections for individuals from a variety of groups to include color, national origin, sex, religion, disability, language, etc. As universities, states, and local governments have extended protections to more individuals and groups, such as those based on gender identity or expression, sexual orientation, marital status, genetic information, and personal appearance, civil rights and the protections that extrapolate out of court decisions are ever-changing.

The United States once had a society legally segregated by race and gender among other factors with no attempts to protect individual rights or access to Extension programs until the Civil Rights Acts of 1866 and 1875 (figure 5). Historically the Land-Grant Acts of 1862 and 1890 created institutions for different races that were based on the laws of that particular time. With the change in population, the efforts to incorporate our programs moved from integration initiatives as a result of the Civil Rights Act of 1964, to affirmative action programs utilized as legal tools of integration, to diversity, and now to diversity, equity, and inclusion programs that are expected within Extension operations. The current conversations, program focus, and individual and institutional efforts have added a new focus of inclusion with equity and equal access measurements. The ultimate challenge and opportunity for Extension educators, agents, specialists, paraprofessionals, staff, executives, and managers are assuring programs are not only diverse

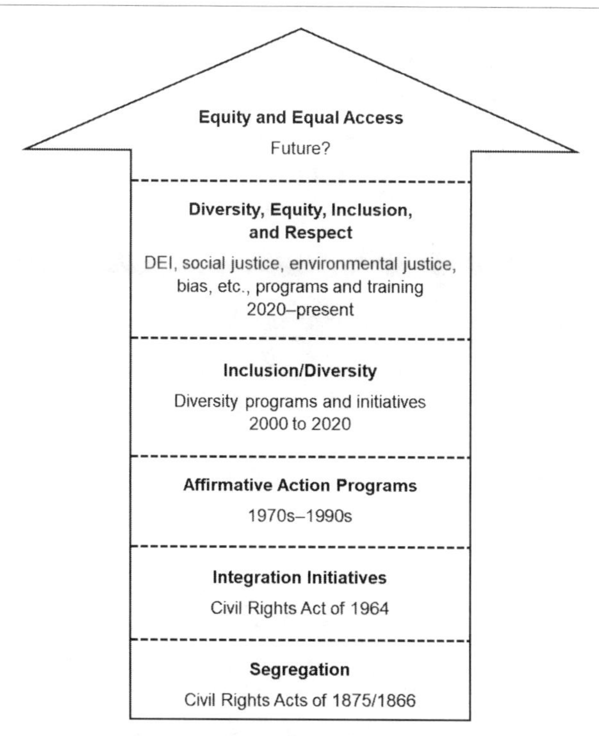

FIGURE 5. The Evolution of Characteristics. Race, ethnicity, color, sex, and diversity variations. *Norman E. Pruitt and Latoya M. Hicks.*

but are truly inclusive of all individuals and provide opportunities to access program benefits and services. Are we in Extension ready to take bold steps such as NASCAR:

> Like many organizations, the National Association for Stock Car Auto Racing (NASCAR) had big choices to make this summer beyond the typical business decisions. NASCAR leaders had to manage safety precautions or holding races during the COVID-19 pandemic and respond to public outcry for social justice following the killing of George Floyd. (Nagele-Piazza, 2020)

Conclusion

This chapter presented the civil rights legal and regulatory compliance environment of the land-grants (figures 1 through 3), the ultimate program beneficiaries (individuals/groups) derived from general population data (figure 4), and the concepts of underserved, underrepresented, and socially disadvantaged audiences. Lastly, we shared the evolution of civil rights laws as our nation's population has changed since the initial recording of race data in 1610 and the first U.S. Census of 1790.

We identified some key tools and resources for the Extension administrator, agent/educator, staff, and volunteer to utilize along the journey to equal access, social justice, acceptance, and full access to all individuals from various walks of life. The laws and regulations presented are the minimum standards and thresholds derived from such regulations, government guidance, affirmative action plans, and diversity programs. These instruments may or may not yield full equity, equal access, and/or inclusion. Such will require a collective effort and vision to move beyond these minimum standards and require a more in-depth review of Extension programs rather than documenting all reasonable efforts or just reaching program parity. We must also be prepared to measure how we move the needle of diversity, equity, and inclusion while assuring respect for all. Data, training, surveys (climate), and clear DEI plans are crucial tools along the path of providing equal access to programs and employment.

As illustrated in figure 5, how and when will Extension reach full inclusion with equity and equal access? What data will be necessary to measure our efforts and identify our successes beyond parity? How will the land-grant environment continue its public obligation to provide benefits, services, and employment to all individuals? We leave the reader with these questions to begin a conversation within their Extension organization on reaching mindsets beyond the minimum standards to the ultimate goal of equity and inclusion as a shared vision of Extension evaluating programs and employment delivery on a nondiscriminatory basis.

NOTE

1. "*Underserved* means an area or community lacking an adequate level or quality of service in an eligible program, including areas of duplication of service provided by an existing provider where such provider has not provided or will not provide adequate level or quality of service" (SUTA, 2012). "Underserved—The term 'underserved population' means a population of individuals, including urban minorities, who have historically been outside the purview of arts and humanities programs due to factors such as a high incidence of income below the poverty line or to geographic isolation" (OMB, 2014, p. 701).

   "Underrepresented minority means any ethnic group—African American, Alaskan Native, American Indian, Asian-American, Hispanic American, Native Hawaiian, Pacific Islander, or any other group—whose representation among food and agricultural professionals in science, technology, engineering, and mathematics (STEM) fields is disproportionately less than their proportion in the general population as indicated in standard statistical references, or as documented on a case-by-case basis by national survey data submitted to and accepted by the Secretary" (NIFA, 2017, p. 4).

   "Socially-Disadvantaged Farmer or Rancher (SDA)—A farmer or rancher who is a member of a group whose members have been subjected to racial or ethnic (and in some cases gender) prejudice because of his or her identity as a member of the group. The definition of SDA farmers varies by Title within the farm bill; some titles include gender and some are limited to racial or ethnic groups" (NSAC, 2019).

   "Socially disadvantaged farmer or rancher—. . . a farmer or rancher who is a member of a socially disadvantaged group" and "Socially disadvantaged group—. . . a group whose members have been subjected to racial or ethnic prejudice because of their identity as members of a group without regard to their individual qualities" (FACT, 1990, pp. 128–129).

REFERENCES

Agricultural Research, Extension, and Teaching. (2021). 7 USC Chapter 64: §3102. Purposes of Agriculture, Research and Extension. http://uscode.house.gov/.

Americans with Disabilities Act. (1990). Public Law 110-325, as amended 2008. Section 12182. Prohibition of Discrimination by Public Accommodations. https://www.ada.gov/pubs/adastatute08.htm.

Bezrukova, K., Jehn, K. A., & Spell, C. S. (2012). Reviewing Diversity Training: Where We Have Been and Where We Should Go. *Academy of Management Learning & Education*, 11(2), 207–217.

CFRDA (Consolidated Farm and Rural Development Act). (1992). Public Law 102-554. Section 21, Equal Access to FmHA Assistance by Gender. https://uscode.house.gov/statutes/pl/102/554.pdf.

Civil Rights Act. (1964). Public Law 88-352 [H.R. 7152].-Nondiscrimination in Federally Assisted Programs. https://www.govinfo.gov/content/pkg/STATUTE-78/pdf/STATUTE-78-Pg241.pdf.

Civil Rights Restoration Act. (1987). Public Law 100-259 [S. 557]. http://uscode.house.gov/statutes/pl/100/259.pdf.

College of Agriculture and Natural Resources. (2021). Diversity, Equity, Inclusion, & Respect Plan. https://agnr.umd.edu/about/diversity-and-inclusion.

DOJ (U.S. Department of Justice). (2019a). Limited English Proficiency Maps. https://www.lep.gov/maps/.

———. (2019b). Overview of Title VI of the Civil Rights Act of 1964. https://www.justice.gov/crt/fcs/TitleVI-Overview.

———. (2020). Title VI Legal Manual. Civil Rights Division. https://www.justice.gov/crt/book/file/1364106/download.

DOL (U.S. Department of Labor). (2019). Executive Order 11246 as amended. Part II—Nondiscrimination in Employment by Government Contractors and Subcontractors. Office of Federal Contract Compliance Programs. https://www.dol.gov/ofccp/regs/statutes/eo11246.htm.

EdSource. (2021). Teachers Reflect on a Year of Covid: Students Struggling, Others Thriving. https://edsource.org/2021/teachers-reflect-on-a-year-of-covid-students-struggling-others-thriving/649705.

Ewoh, A. (2013). Managing and Valuing Diversity: Changes to Public Managers in the 21st Century. *Public Personnel Management*, 42(2), 107–114.

FACT (Food, Agriculture, Conservation, and Trade Act). (1990). Public Law 101-624. https://www.agriculture.senate.gov/imo/media/doc/101-624.pdf.

Food Stamp Act. (1977). Public Law 108-269, July 2004. Administration, Sec. 11 [7 U.S. C. 2020], (c). https://www.fns.usda.gov/snap/food-stamp-act-1977-pl-88-525a.

GAO (U.S. Government Accountability Office). (1981). Cooperative Extension Service's Mission and Federal Role Need Congressional Clarification. August 21. https://www.gao.gov/products/CED-81-119.

Gibson, C., & Jung, K. (2002). Historical Census Statistics on Population Totals by Race, 1790 to 1990 and by Hispanic Origin 1970 to 1990, for the United States, Regions, Divisions and States. Table 1 "United States–Race and Hispanic Origin: 1790 to 1990." https://census.gov/content/dam/Census/library/working-papers/2002/demo/POP-twps0056.pdf.

Jones, C. (2006). Falling between the Cracks: What Diversity Means for Black Women in Higher Education. *Policy Futures in Education*, 4(2), 145–159.

Lempinen, E. (2021). Across America, Trust Is Falling Apart. Can It Be Rebuilt? *Berkeley News*, January 14. https://news.berkeley.edu/2021/01/14/our-national-trust-is-falling-apart-can-it-be-rebuilt/.

Leonard, B. (2012). Higher Education Needed for Most Future Jobs: SHRM Study. Society for Human Resource Management, October 12. https://www.shrm.org/hr-today/news/hr-news/pages/higher-education-needed.aspx.

LLI (Legal Information Institute). (2019). Code of Federal Regulations. Title 7: Agriculture, Subpart

A—Nondiscrimination in Federally-Assisted Programs of the Department of Agriculture. https://www.law.cornell.edu/cfr/text/7/part-15/subpart-A.

———. (2019b). Code of Federal Regulations. Title 7: Agriculture, Subpart D—Nondiscrimination in Federally-Assisted Programs of the Department of Agriculture. https://www.law.cornell.edu/cfr/text/7/part-15a/subpart-D.

Nagele-Piazza, L. (2020). NASCAR Takes Bold Action to Combat Social Injustice. Society for Human Resource Management, October 22. https://www.shrm.org/resourcesandtools/legal-and-compliance/employment-law/pages/nascar-takes-bold-action-to-combat-social-injustice.aspx.

NARETPA (National Agricultural Research, Extension, and Teaching Policy Act). (1977). Public Law 107-293. https://nifa.usda.gov.

NIFA (National Institute of Food and Agriculture). (2017). Women and Minorities in Science, Technology, Engineering, and Mathematics Fields Program. https://nifa.usda.gov/sites/default/files/resources/FY%2017%20WAMS%20FAQ.pdf.

———. (2019). "And Justice for All" poster. https://nifa.usda.gov/resource/and-justice-all-poster.

NSAC (National Sustainable Agriculture Coalition). (2019). Glossary and Acronyms. http://sustainableagriculture.net/publications/grassrootsguide/glossary/.

OMB (Office of Management and Budget). (2014). Executive Office of the President of the United States, Title IV–General Provisions. Sec. 413. (b) (1).

Rehabilitation Act. (1973). Public Law 93-112 [H.R. 8070]. https://www.govinfo.gov/content/pkg/STATUTE-87/pdf/STATUTE-87-Pg355.pdf.

———. (1973b). Public Law 93-112 [H.R. 8070]. Electronic and Information Technology. https://www.section508.gov/manage/laws-and-policies/#508-policy.

SUTA (Substantially Underserved Trust Areas). (2012). 7 CFR 1700, Subpart D, §1700.101 Definitions. Rural Utilities Services, Federal Register. https://www.federalregister.gov/documents/2011/10/14/2011-26133/substantially-underserved-trust-areas-suta#p-69.

UMD (University of Maryland College Park). (2019). Non-discrimination Policy and Procedures. Policy VI-1.00(B). https://president.umd.edu/administration/policies/section-vi-general-administration/vi-100b.

———. (2021). Faculty Advancement at Maryland for Inclusive Learning and Excellence (FAMILE). https://faculty.umd.edu/main/appointments/faculty-hiring-process#famile-faculty-advancement-at-maryland-for-inclusive-learning-and-excellence.

U.S. Census. (2019). Race Defined. https://www.census.gov/topics/population/race/about.html.

———. (2020a). Decennial Census. https://www.census.gov/en.html.

———. (2020b). Population in the Colonial and Continental Periods: A Century of Population Growth. https://www.census.gov/history/pdf/colonialbostonpops.pdf.

USDA (U.S. Department of Agriculture). (1999). Departmental Regulation Number 4330-002. "Nondiscrimination in Programs and Activities Receiving Federal Financial Assistance from USDA."

———. (2017). Maryland Census of Agriculture. National Agriculture Statistical Service, Table 63. https://www.nass.usda.gov/AgCensus/.

USDA (U.S. Department of Agriculture) Extension Service. (1983). "Plans and Procedures for Program Participation Data and Information Collection in the Cooperative Extension Service."

Culture and Culturally Relevant Programming

Nia Imani Fields and Fe Moncloa

Culture is a complex concept, yet it is often oversimplified. In fact, some only associate culture with nondominant identities.[1] Dominant identities represent the values, practices, languages, and traditions that are assumed to be the most accepted and influential within a given society (Abrams & Hogg, 1990; Hogg et al., 1995). These dominant cultural identities, then, tend not to be seen as cultural at all and instead as the norm. In American culture, dominant identities include white, middle-class/wealthy, heterosexual, male, Christian, people with college degrees, people without disabilities, American citizens, and English-speaking people. Culture is interwoven among and between each of these identities as it is with *all* people.

The truth of the matter is, we are all connected to and influenced by multiple aspects of culture. Many people describe culture in its objective form—the things that we can see. These aspects are explicit forms of culture and may include traits such as dance, art, music, customs, clothing, language, and sports. If we dive deeper into understanding our own culture as well as other cultures around us, we begin to recognize subjective forms of culture—the things we may not see. These aspects may include roles, myths, beliefs, social expectations, and values. You see, culture describes the shared experiences of people, including their languages, values, customs, and worldviews. Culture influences how people meet their basic human needs, how they learn and understand the world, solve problems, and communicate, among other things.

Culture is dynamic and fluid. At the core, we have an individual culture often passed down through our families from generation to generation. These traditions may influence what food we eat, our style of clothing, or how we raise our children. We are also connected to and influenced by organizational culture within the institutions and workplaces in which we are situated. This often sets the tone for things like the appropriate attire, whether we eat lunch together or alone in our office spaces, and how policy decisions are made. Organizational culture may also dictate who has power and who does not, and who feels welcome in our programs and who does not.

Organizational culture does not change as easily or frequently as our individual cultural identities. Furthermore, organizations that were formed generations ago for the purpose of serving a homogeneous audience often have a harder time changing the culture to be more inclusive of increasingly diverse communities.[2] Extension is one such organization that was designed with and for the white dominant culture—in this case, rural white Americans. The Smith-Lever Act formalized Extension in 1914, establishing a partnership between the U.S. Department of Agriculture (USDA) and land-grant universities to, in essence, extend a bridge between land-grants and communities. Extension became a local, state, and national system to apply university research through community education in agriculture.

> At that time, more than 50 percent of the U.S. population lived in rural areas, and 30 percent of the workforce was engaged in farming. . . . Extension has [since] adapted to changing times and landscapes. Fewer than 2 percent of Americans farm for a living today, and only 17 percent of Americans now live in rural areas. (National Institute of Food and Agriculture, 2020)

Extension is now situated within urban, suburban, and rural communities to engage in grassroots community education ranging from youth development to climate change, and from food security to environmental justice. However, Extension is still widely led, designed, and staffed by dominant cultures. In fact, some argue that Extension doesn't currently have a state program that can claim to have a sustainable and effective culturally diverse organization—both internally and among the communities we engage (Schauber, 2001). It is important to acknowledge and be informed by the history of this organization as it undergirds the current culture and legacy of Extension.

In both the introduction and the opening chapter of this book, the authors discuss Extension's origins and the disparities within. They describe the segregated Extension story and are a reminder that the land-grant systems, from the institutions to the land they reside on, are all deeply rooted in racism, sexism, and colonialism. It is important to understand Extension's story and how its legacy perpetuates dominant privilege within the organizational culture today.

In addition to individual and organizational forms of culture, there is a broader community and societal culture that also shapes who we are and what we value within society. Communities are rich in culture and should be the driving force behind Extension and community education programs. Culture is, after all, at the core of how we view the world and make meaning of information. It is, therefore, a fundamental element of learning and community education.

Culturally Relevant Teaching

The community work of Extension is most genuine and impactful when it is culturally relevant. Culturally responsive teaching uses the cultural knowledge, lived experiences, frames of references, and values of people to make the learning experience more meaningful and effective (Gay, 2010). Culturally responsive teaching is validating and affirming because it acknowledges the legitimacy of diverse identities; it builds bridges between individual cultural experiences, academic abstractions, and sociocultural realities; it adapts teaching to connect to different learning styles; it teaches learners to explore and appreciate their own culture; and it incorporates diverse voices, resources, and materials in the learning experience (Gay, 2010).

Ladson-Billings challenges educators to think even beyond the learning experience and to consider the true purpose of education. In its most equitable and authentic form, education should be guided by a culturally relevant pedagogy that empowers communities intellectually, socially, emotionally, and politically to enact change for a more just society (Ladson-Billings, 1994). Culturally relevant pedagogy is "a pedagogy of opposition [that is] committed to collective, not merely individual, empowerment" (Ladson-Billings, 1995, p. 160). This pedagogy rests on three criteria: (1) students must experience academic success; (2) students must develop and/or maintain cultural competence; and (3) students must develop a critical consciousness through which they challenge the status quo of the current social order (Ladson-Billings, 1995, p. 160).

We acknowledge that Extension is not where it needs to be in terms of organizational changes needed to foster a culture of relevant programs that are authentically guided by the diverse voices we hope to engage. However, there are some examples within Extension where communities and educators (agents) together have been engaged in culturally relevant experiences that have led to thriving communities, increased civic engagement, and movements of environmental, health, and food justice. Some examples can be found in the subsequent chapters of this book. The following case study is an example of a program where the educator employs culturally relevant teaching to engage young immigrants in a youth development program.

Case Study

In a large city in California, an Extension agent who is a native Spanish speaker prepares for her first three-hour session with twenty-four Spanish-speaking immigrant Latinx youths. The goal of the project is to build the capacity of young people to both learn computer science and teach computational thinking skills to younger children. The educator relies on her past experience of cultivating people's agency to serve as teen teachers and on her recent experience of coaching teens from various 4-H clubs to teach computer science.[3] But this setting is different: these Latinx youth only speak and write low-literacy Spanish, they are new immigrants to the United States, and they have limited interaction with technology. Of the twenty-four students in the class, five own cell phones and none of them have access to computers or Wi-Fi at home, and they do not use computers at school.

To facilitate youth engagement in this project, all activities are conducted during school hours and healthy snacks are abundant. Transportation is a known barrier to participation for Latinx youth (Erbstein & Fabionar, 2014), and by implementing a school-based project, there is a likelihood that attendance will be consistent. Nevertheless, the first day two students are absent because they have a meeting with immigration lawyers.

Without knowledge of who her students will be, except for the known facts mentioned earlier, the Extension agent prepares three mindfulness activities, two icebreakers, and two ways of teaching "how to teach younger children." The thinking behind this preparation is that not all activities will resonate with youth, and the agent will not know this until she starts teaching. In fact, she is aware that she may have to create something on the spot in response to the Latinx youths' needs. To facilitate learning, the educator develops a PowerPoint presentation where images of Latinx adolescents are sprinkled throughout, and all activities include step-by-step instructions in simple Spanish.

As she prepares these activities, she reflects on her privilege as an educated Latinx, her memories as an immigrant to the United States, the wealth of resources she had as a teenager compared to her future students, and the social capital she had access to in high school. The educator acknowledges that while she is Latina, her cultural values reflect those of a person that has lived in the United States her adult life, and she acknowledges her cultural values may be similar to and different from those of her students. The educator anticipates that she and her students will share a collectivist culture with strong family values, *familismo*, a hierarchical culture that values *respeto* (respect), and a tendency to emphasize relationships more than tasks.

Naming the Moment

On the day of the training, the Extension educator wears business casual attire. In Latinx culture the teacher is a well-respected professional and serves as a role model for the next generation. As students arrive to the classroom, the educator welcomes them, introduces herself, and asks for their name. Once everyone is sitting down in their self-selected groups, the session starts. The educator introduces herself; she shares her background, country of origin, and favorite hobby. She asks students how they would like to address her. They all say *maestra* (teacher), and one young person explains that this is done out of respect for her. Before she begins the training, the Extension agent "names the moment" by acknowledging that these are difficult times in the United States for immigrant families, and that she and others in this school and community are committed to supporting their families' and the youths' safety. By "naming the moment" the educator acknowledges the socioeconomic and political context that shapes youths' and their families' lives and creates a space for future conversations on these topics.

The first activity is a mindfulness activity that aims to bring students into the present, amidst all their worries. Students are guided through a meditative activity to try to help them put their daily worries about immigration, access to food, and limited English knowledge aside. They are guided through an adaptation of the hot air balloon ride. While the hot air balloon guided meditation allows youths to view possible solutions to their issues, this activity is adapted with the purpose of suspending youths' everyday concerns for a moment, since as new immigrants they have limited power over discrimination and poverty. The teacher asks students to close their eyes and to trust her as she guides them through images.

The second activity is designed to get to know students' present and future interests. It also serves to help students critically reflect on their own lives and the society they live in. It is an adaptation of the shield activity (Miner et al., 2015). The adaptation aims to validate students' cultural emphasis on family. This is an individual or group activity where young people draw or describe their answers to the following questions: What are the things that you enjoy doing with your friends or family? If money was not a barrier, what things would you like to do with your friends or family? In addition to your family, who are the people that matter to you the most? What are the activities that matter to you? What are the local, state, national, or international issues or causes that matter to you that need to be addressed? This activity serves as both an icebreaker and an assessment of students' existing strengths and their challenges. Some students organize their thoughts before answering the questions, others dive in and run out of space, some youth draw their answers, others do a combination. Some students write well in Spanish, while others need assistance to spell words, and some need assistance to write. They all write "spoken" Spanish. As students share their stories, the educator learns from them their cultural values, their passions and interests, and their willingness to learn. In this exercise, the Extension agent is practicing culturally responsive teaching and uses "the cultural knowledge, prior experiences, frames of reference, and performance styles of ethnically diverse students to make learning encounters more relevant to and effective for them" (Gay, 2010, p. 31).

During the activity, the educator learned that in addition to teaching youth how to teach computer science to younger children, she will be teaching Spanish and English, the similarities and differences between American cultural traditions and the youths' culture, university or career culture, and Silicon Valley techie culture. The educator will seek the assistance of people who are engulfed in these cultures to share their experience with youth. In culturally relevant teaching, the onus of educating youth is on the teacher (Gay, 2013). The focus of this project goes beyond computer science education; it aims to develop youths' global identities as it seeks to embrace language, arts, and science (Ladson-Billings, 2006).

At the conclusion of this activity, trust is beginning to be built among students and the Extension agent. The Extension agent waited until the youths were relaxed and more familiar with the teacher to communicate the expectations for participation. She communicates high expectations for each student and for teamwork, and her commitment to them to ensure their success. The educator adds that the computer science project will serve to honor young people's cultural strengths while facilitating their acquisitions of computer science knowledge to provide access to a wider culture, and possible access to a chance to improve their socioeconomic status.

Learning Stages of Child Development

The purpose of the next activity is for teens to learn the characteristics of younger children. Youths who have siblings or cousins who are nine- to twelve-year-olds are asked to describe their interactions with children. At first no one speaks. The educator asks the youths to remember the last birthday party they went to, and what games children played. The conversation starts to flow. In Latin American culture, children's birthday parties are intergenerational. Parents, aunts, grandmothers, siblings and their friends, and friends of friends, plus the birthday child and his/her friends attend. Stories of short-lived games and music, plus the food, describe the birthday party. From these stories the educator introduces the characteristics of children ages nine through twelve. As she presents each characteristic, she asks teens whether they have seen it displayed in their interactions with children, and to share. Next, young people work in self-selected groups to figure out how they would teach a color-mapping activity to younger children.

While teens are engaged in activity, one teen brings out her cell phone and turns the music on. The educator asks students to pause their work for a second and asks them if they like working with music. They all say yes. Next, she facilitates an impromptu set of guidelines with the youths on the type of music, the volume, and where to place the cellphone in the room. All guidelines are generated by the youths. As a result, bringing music into the classroom became the norm for this cultural group of beginner computer scientists.

We Are Computer Scientists!

In preparation for the second three-hour session, the educator reviews the written word of the students to ascertain their literacy levels in Spanish and reflects on the next steps for the project. One of the challenges faced by the educator is that most of the computer science curriculum is in English. In addition to preparing to teach the session, she must translate all documents into simple Spanish. However, language is not the only barrier. It seems that these young people have had limited experience in developing their critical thinking skills, have difficulty following directions, and have a limited attention span.

She reflects on the original plan for this project. The intent was that teens would learn a variety of computational thinking skill activities, and then teach these activities to the same grade. It is evident to the educator that she cannot follow this plan and set up the teens for success. Her reflection on what is the best course of action takes into account the youths' commitment to learning, willingness to contribute to children's learning, and desire to be computer scientists.

Based on these strengths and constraints, and in response to the Latinx teens' needs, the educator changes her original plan and asks the youths to self-select into three teams. Each team will become an expert in one computer science activity and will teach this activity once a week to children in different grades. The educator proceeds and teaches the youths three computational thinking activities. For each activity, the educator first elicits a personal experience from the youths that can be linked to the activity, then teaches the vocabulary for the session, tying it back to personal experience, and then teaches the activity.

At the conclusion of each activity, the educator debriefs with students how each activity was taught. She asks students: How did I engage you in the activity? What questions did I ask? How did I build on what you already know? And so forth. As she does this, she is reteaching important engagement elements of each activity. During the first debrief, she notices that students do not take notes on strategies to engage children and asks the students why. One student responds that in general, they only take notes when the teacher tells them to. Other students agree. The educator pauses the conversation and shares with students an impromptu lesson on note-taking and why it matters, and how it can improve their learning. During this quick teach, note-taking is also presented as a skill that may help them in their careers.

At this intersection, the educator is torn between stepping into the expected role that students have of teachers—who tell them when to take notes and what to write—or to cultivate their agency by encouraging them to take notes on their own of what they think is most important to them. She decides to start with the expected role and to slowly cultivate the youths' intrinsic motivation to take notes. By starting with the expected role, the educator is meeting Latinx youth where they are at and is responding to their immediate needs.

After all activities are taught and debriefed, student teams prepare to teach each other for the first time. The instructor gives each team a PowerPoint deck with their activity, worksheet, and materials. They are encouraged to change the PowerPoint to meet their needs. Each team learns to self-evaluate after teaching, focusing on the good, the better, and the how. All teams acknowledge they need more preparation time to be more organized.

A week later, the teens are ready to teach to younger children. One team walks into the fourth-grade classroom, and the teacher welcomes the computer scientists into the class and states: "we are computer scientists." The stage is set for instruction to begin.

Ten Tips for Creating Culturally Relevant Programs

To begin to learn how to adapt one's educational program to ensure it is culturally relevant, Extension professionals need to engage in the life-long journey of increasing their intercultural competence, defined as a "set of cognitive, affective and behavioral skills and characteristics that support effective and appropriate interaction in a variety of cultural contexts" (Bennett, 2008, p. 97). The first five tips listed aim to support an individual's intercultural competence development.

Intentional self-reflection to understand one's cultural norms, values, beliefs and behaviors. Before learning about other cultures, it is imperative to engage in an intentional self-reflective process "focused on understanding patterns of difference and commonality between yourself (and your cultural group) and other cultural groups perceptions, values and practices" (Hammer, 2012, p. 29). Understanding how a person's upbringing and experiences throughout life have influenced these cultural patterns is a first step. Extension educators and administrators need to have a clear understanding of their ethnic and cultural identities.

Experience cultures different from your own. Experiencing other cultures can take on many forms such as watching a movie, traveling abroad, engaging in conversations with people, or participating in trainings, to name a few. In learning and experiencing cultures different from one's own, it is important to engage in self-reflection on the similarities and differences between one's own culture and other people's cultures.

Develop an appreciation and respect for diverse cultural beliefs and values, beyond objective surface understanding of culture, toward a deeper subjective understanding. A key practice in culturally relevant teaching is to accept other ways of knowing and doing, and to acknowledge how these behaviors may be similar to or different from one's own culture. In this reflection, it is important to withhold judgment. In this process, educators begin to understand various cultures and cease "othering" cultures different than one's own. This is more easily said than done.

Evaluate overgeneralizations and stereotypes. Seek clarification when needed. A challenge in working with diverse populations from various cultural backgrounds is being able to apply cultural generalizations appropriately. Cultural generalizations are flexible and serve to categorize members of a cultural group as having similar shared behaviors. However, there are always variations among cultural subgroups that are often influenced by social contexts. This challenge can be addressed by understanding the variations among cultural subgroups and taking into account the social-economic and political contexts that influence the lives of youth, families, and communities.

Stereotypes, on the other hand, are generalizations that are used to describe members of a cultural group as having the same cultural characteristics. In general, stereotypes tend to describe cultural groups in negative terms. Resist using stereotypes to describe cultural similarities among cultural groups, and consider checking for understanding and clarification from the cultural group(s) one is working with.

Use materials that reflect people, language, art, music, stories, and games from various cultural traditions. In the teaching or engagement of diverse populations, it is imperative to include images, poems, art, and stories of various cultural groups in the handouts, materials, or presentations. Music can be included before the presentation and while participants engage in group work. Examples of art, stories, and games in one's teaching communicates an appreciation for various cultural groups. In addition, the material needs to be relevant to the audience as well.

Incorporating diverse perspectives in teaching materials, with the content adapted to meet the interests of participants, will allow participants to realize that Extension educators value and appreciate diversity.[4]

Demonstrate to participants that you care, and provide experiences that facilitate engagement and discussion of their own cultural backgrounds and assets. Educators shall seek opportunities to elicit participants' experiences to learn what they know and learn about their cultural backgrounds. These activities may include engaging diverse populations in the planning, implementation, and/or evaluation of Extension programs, or creating interactive activities where participants share their cultural assets. Educators can demonstrate caring attitudes toward diverse learners by "naming the moment" and sharing their concerns, hopes, and dreams.

Communicate high expectations for diverse participants. It is important to communicate to diverse learners the high expectation for participating in Extension programs. Everyone can learn when the material is taught in a culturally relevant manner.

Incorporate multiple assessment tools. In most Extension workshops or courses participants complete a pre-post test to evaluate knowledge, skills, and attitudes. However, this assessment tool may not be appropriate for diverse cultural groups. Evaluation must be rooted in cultural relevance, social justice, deliberative democracy, participatory research, and empowerment research to amplify the voices of underrepresented communities. Examples of alternative assessments include portfolios, testimonies, or group interviews.

Ensure practices, guidelines and policies are created or adapted with diverse populations to be more inclusive. By engaging diverse populations in the planning, implementation, and evaluation of programs and policies, Extension can develop or adapt their practices to be more inclusive.

Advocate for systemic organizational change to respond to the needs and interests of diverse populations. To build and sustain a culturally competent organization, Extension needs to advocate for the allocation of resources for staff development in intercultural competence. An interculturally competent Extension organization will facilitate equity in policies and practices, in hiring practices and evaluation of personnel, and in shared leadership and power among leaders from diverse cultural backgrounds. Advocating for systemic organizational change includes evaluating the organization's cultural competence regularly and conducting an organizational equity assessment.

Conclusion

Educators have an immense role in the education and development of youth and communities. It is important to note that developing programs that are not culturally relevant is in fact a form of injustice as it silences and ignores the myriad cultures that exist within our society. Extension programs and other forms of community education must authentically engage communities and diverse audiences to ensure the community is in fact at the center of our work.

As educators reflect on their role in and responsibility to culturally relevant teaching, there are some questions that can be asked:

- In what ways does your program make the community's identity central to the program's goals and activities?
- How are you identifying and including diverse voices of history, theories, and experiences in your curriculum and teaching materials?
- Through what process(es) are you recognizing, identifying, and reacting to learning environments that are predominately and historically filled with a dominant culture?
- What learning methods do you use to develop the critical awareness of your audience?
- In what ways does your program foster empathy?
- Who defines what success looks like after an educational or community engagement experience?

Ongoing reflection and relearning is a critical component of fostering culturally relevant experiences with communities. If Extension intentionally works with and for the people within community, we stand a chance to address relevant concerns and equitably solve problems together.

NOTES

1. Culture: The shared experiences of people, including their languages, values, customs, beliefs, and more. It also includes worldviews, ways of knowing, and ways of communicating. Culture is dynamic, fluid, and reciprocal. Elements of culture are passed on from generation to generation, but culture also changes from one generation to the next (American Evaluation Association, 2011; Deen et al., 2015).
2. Inclusion: A state of being valued, respected and supported. Inclusion authentically puts the concept and practice of diversity into action by creating an equitable environment where the richness of ideas, backgrounds, and perspectives is harnessed (Jordan, 2011; Baltimore Racial Justice Action, 2016).

3. Agency: The belief that you have the skills, resources, and power to make a difference, particularly as it relates to social justice.
4. Diversity: Differences among people with respect to age, socioeconomic status, ethnicity, gender, physical and mental ability, race, sexual orientation, spiritual practices, and other human differences (Deen et al., 2015).

REFERENCES

Abrams, D., & Hogg, M. A. (1990). *Social Identity Theory: Constructive and Critical Advances.* London: Harvester/Wheatsheaf.

American Evaluation Association. (2011). American Evaluation Association Statement on Cultural Competence in Evaluation. Approved April 22. https://www.eval.org/About/Competencies-Standards/Cutural-Competence-Statement.

Baltimore Racial Justice Action. (2016). Our Definitions. November 30. http://bmoreantiracist.org/resources/our-definitions/.

Bennett, J. M. (2008). Transformative Training: Designing Programs for Culture Learning. In M. A. Moodian (Ed.), *Contemporary Leadership and Intercultural Competence: Understanding and Utilizing Cultural Diversity to Build Successful Organizations* (95–110). Thousand Oaks, CA: Sage.

Deen, M., Parker, L., & Huskey, M. (2015). *Navigating Difference Cultural Awareness Notebook.* Pullman, WA: WSU Extension Publishing.

Erbstein, N., & Fabionar, J. (2014). Latin@ Youth Participation in Youth Development Programs. University of California Division of Agriculture and Natural Resources, September 24. http://cesantaclara.ucanr.edu/files/261436.pdf.

Gay, G. (2010). *Culturally Responsive Teaching: Theory, Research and Practice* (2nd ed.). New York: Teachers College Press.

———. (2013). Teaching To and Through Cultural Diversity. *Curriculum Inquiry*, 43(1), 48–70.

Hammer, M. R. (2012). *A Resource Guide for Effectively Using the Intercultural Development Inventory (IDI).* Berlin, MD: IDI, LLC.

Hogg, M., Terry, D., & White, K. (1995). A Tale of Two Theories: A Critical Comparison of Identity Theory with Social Identity Theory. *Social Psychology Quarterly*, 58(4), 255–269.

Jordan, T. H. (2011). Moving from Diversity to Inclusion. Profiles in Diversity, March 22. http://www.diversityjournal. com/1471-moving-from-diversity-to inclusion/.

Ladson-Billings, G. (1994). *The Dreamkeepers.* San Francisco: Jossey-Bass Publishing Co.

———. (1995). But That's Just Good Teaching! The Case for Culturally Relevant Pedagogy. *Theory into Practice*, 34(3), 160–165.

———. (2006). From the Achievement Gap to the Education Debt: Understanding Achievement in U.S. Schools. *Educational Researcher*, 35(7), 3–12.

Miner, G., Iaccopucci, A., & Trzesniewski, K. (2015). *iChampion 4: Healthy and Thriving Adult Volunteer Leader Guide* (2nd ed.). Davis: University of California Division of Agriculture and Natural Resources.

National Institute of Food and Agriculture. (2020) Cooperative Extension History. https://nifa.usda.gov/cooperative-extension-history.

Schauber, A. (2001). Effecting Extension Organizational Change toward Cultural Diversity: A Conceptual Framework. *Journal of Extension*, 39(3), 3FEA1. https://archives.joe.org/joe/2001june/a1.php.

Cultural Competence in Evaluation: One Size Does Not Fit All

Megan H. Owens, Michelle Krehbiel, and Teresa McCoy

> Evaluators should not ignore imbalances of power or pretend that dialogue about evaluation is open when it is not. To do so is to endorse the existing social and power arrangements implicitly and to evade professional responsibility (House & Howe, 2000, pp. 9–10).

The tenets of cultural competence, social justice, deliberative democratic evaluation, participatory evaluation, empowerment evaluation, and other approaches ensure that underrepresented groups' voices, opinions, and realities are at the forefront of evaluation scholarship and practice. Discussion of these topics has been prominent within the major evaluation professional organization in the United States—the American Evaluation Association (AEA)—and in its official journals for several decades (Azzam & Levine, 2014; Bledsoe, 2014; Fitzpatrick, 2012; Madison, 2007; McBride, 2011; SenGupta, Hopson & Thompson-Robinson, 2004). In 2011, AEA issued an official "Statement on Cultural Competence in Evaluation" that defined cultural competence as both a stance and a process that evaluators engage in rather than as a set of fixed knowledge or skills (AEA, 2011, p. 1). In that statement, AEA emphasized three reasons why cultural competence is important:

1. Evaluators have "to be culturally competent in order to produce work that is honest, accurate, respectful of stakeholders, and considerate of the general public welfare."
2. Valid evaluation results require that "diverse voices and perspectives are honestly and fairly represented."
3. Evaluation theories are products of the cultures in which they are developed and have to be "scrutinized . . . in order to understand how they describe societal issues and how to address them." (AEA, 2011, p. 1)

In 2018, AEA revised and updated its "Guiding Principles for Evaluators" in recognition of the rapid change in the world and the need for evaluators to be culturally competent. The principles seek to "govern the behavior of evaluators in all stages of the evaluation" and encompass five areas:

1. Systematic Inquiry: Evaluators should conduct data-based studies that are thorough, methodical, and contextually relevant. In addition, evaluators should ensure that stakeholders understand the preferred methods and are clear about limitations.

2. Competence: Evaluators who provide professional services must have the education, skill set, and abilities to carry out the evaluation in a competent manner. Professional development and training are important to maintaining competence in new methods and approaches.
3. Integrity: Evaluators practice integrity through transparency and honesty with stakeholders. Conflicts of interest, limitations, and changes in the evaluation should be clearly communicated and explained.
4. Respect for People: Evaluators should fairly represent various stakeholder interests and operate ethically through consent, confidentiality, minimizing harm, and maximizing benefits. These principles are outlined in the "Belmont Report," which outlines ethical principles for conducting human subjects research.
5. Common Good and Equity: Evaluators have the responsibility to contextually and culturally consider their position and power within the evaluative context. Evaluation results should "contribute to the common good and advancement of an equitable and just society." (AEA, 2018, p. 4)

Recognizing and Addressing Historical Issues

A review of historical research reveals two prominent cases of what can happen when researchers and evaluators do not adhere to the "Guiding Principles for Evaluators" tenets. The first case is the U.S. Public Health Service "Tuskegee Study of Untreated Syphilis in the Negro Male" in Alabama (CDC, 2015). From 1932 to 1972, four hundred men with syphilis were engaged in a study where they were not informed of their medical diagnosis, not provided with the knowledge or opportunity to consent to the study, and, ultimately, denied the proper treatment of penicillin that was commonly provided to white patients (CDC, 2015). Furthermore, these men never learned they had a contagious disease that could be transmitted to others. Instead, the men believed they suffered from "bad blood" (CDC, 2015).

The second case concerns the harvesting of cancer cells from a Black woman, Henrietta Lacks, while she was undergoing treatment for cervical cancer at Johns Hopkins Hospital in Baltimore, Maryland. These cells, called HeLa (named with the first two letters from her first and last names), became and continue to be fundamental to cancer research. Neither Henrietta Lacks nor her family knew the cells were harvested and continuously used for research. The story of the HeLa cells was not fully explored or exposed until Rebecca Skloot (2010) researched and wrote the book *The Immortal Life of Henrietta Lacks*. Finally, in 2013, Johns Hopkins Medicine reached a formal agreement with the Lacks family "that requires scientists to receive permission to use Henrietta Lacks' genetic blueprint, or to use HeLa cells in NIH funded research" (Johns Hopkins Medicine, n.d.).

The abusive treatment of vulnerable populations, as described in these two cases, informed the research regulations developed by the U.S. Department of Health and Human Services (HHS), Office for Human Research Protections, Code of Federal Regulations, Title 45, Part 46. The ethical principles in these regulations were formally presented in "The Belmont Report: Ethical Principles and Guidelines for the Protection of Human Subjects of Research" in 1979, and include respect for persons, beneficence, and justice (HHS, 1979).

The principle of respect for persons means that people must enter into research studies voluntarily and fully informed. Furthermore, there is a recognition that certain groups of people within the U.S. populations may have diminished capacity and lack the opportunity for self-determination. The populations that are traditionally considered vulnerable are minors, pregnant women, prisoners, and persons with cognitive disabilities. However, evaluators should consider current circumstances and ask themselves if there are other populations that might be considered vulnerable or underrepresented. For example, in some areas of the United States, immigration status may constitute a vulnerable population. Geography may create vulnerable groups, such as inner-city youth or rural youth. Evaluators should carefully consider the targeted population, as specific populations may be more reluctant to participate in evaluative studies due to societal, economical, and environmental conditions or hardships.

The second principle, beneficence, means that the research will "do no harm" and will "maximize possible benefits and minimize possible harms" (HHS, 1979). The third principle of justice focuses on determining if certain populations benefit from the research or if certain populations would carry some unjust burden from the research. In the case of the Tuskegee study, it is easy to see how these three ethical principles were not practiced, specifically these men carried an unjust burden during the research project. Extension professionals will encounter these regulations through the application process with university Institutional Review Boards (IRBs). Most often university IRBs require faculty (researchers) to complete Human Subjects Training to be eligible to conduct research. IRBs establish and approve research protocols that help to ensure the safety of participants. While there may be a great degree of complexity and confusion about these IRB reviews, there are good reasons why we adhere to this type of research review (Vo & Archibald, 2018).

Evaluation as Engagement

The ethical principles deriving from the American Evaluation Association as well as the ethical standards of the IRB process are relevant to Extension and community-level evaluators due to the potential role of being "agents of change" who seek to improve society through evaluation (Vo & Archibald, 2018). The perspective of being "agents of change" may appear contrary to the conventional wisdom that evaluators are purely objective observers or simple reporters of program effectiveness.

Since the 1990s, multiple disciplines viewed randomized controlled trials and evidence-based research as the "gold standard" when determining appropriate evaluative methodologies (e.g., Coalition for Evidence-based Policy, 2003; Archibald, 2015). This focus originated in medical research but quickly gained traction across other disciplines such as education or prevention science. The value attributed to objective, evidence-based programs or evaluation emerged from the belief that superior evidence resulted from focusing on theories of change and prioritizing experimental design over other methodologies (Donaldson, Christie & Mark, 2009; Smith, 2012). This focus has been especially true when decision-makers allow "evidence-based" to serve as a determining factor for programmatic and evaluative efforts (Biesta, 2010). Maintaining pure objectivity when delivering an educational program, let alone administering an evaluation within a community one is intimately involved in, is unrealistic (Archibald, 2015). This is especially true for Extension personnel/professionals who must balance leading educational programs with evaluation and research. Community culture, organizational goals, and the Extension professional's desires may influence program and evaluation fidelity (Greene, 2009). Therefore, Extension professionals and evaluators are more likely to be agents of change, working alongside communities rather than serving as objective reporters.

The ability to recognize the broader context surrounding Extension and community-centered programming and evaluation humanizes the role (i.e., agents of change) that evaluators often play in community-level evaluation. Similar to community program providers or local Extension professionals, the role of evaluation can foster change in a community. Data may reveal unmet needs in a community or unexpected program outcomes, which may, and should, ultimately lead to programmatic and community changes. A thorough planning process that considers the role of the evaluator and background of a community is important before collecting data and making judgments (Lofland et al., 2006; Wehipeihana & McKegg, 2018).

Extension personnel must take great care when working with communities or populations considered vulnerable, marginalized, or previously exploited for their position in society (Smalkoski et al., 2016). This sentiment is applicable when designing and conducting evaluation, as the evaluator's position in society, the community, or program may influence decisions related to the evaluation design as well as the information shared by the participants (Archibald, Neubauer & Brookfield, 2018). The evaluator's goals and practices must align with the evaluation project while recognizing important cultural perspectives of the identified community. Symonette, Miller, and Barela (2020, p. 118) argue that "evaluator roles typically confer social power," and urge evaluators to cultivate self-reflection on their practice. This advice is especially important when Extension program personnel are also acting in the role of internal evaluator.

Starting the evaluation process by contemplating a series of general questions will help evaluators situate their position according to AEA and university IRB principles.

- What is the purpose of the evaluation?
- How will the data be used?
- How will the evaluation participants benefit from their engagement?
- Is the evaluator seeking to empower a community to transform?
- Does the evaluator embrace a particular social justice platform?

This practice is important, as the evaluator may not be a member of the community (Wehipeihana & McKegg, 2018). Evaluators must recognize their position and affiliations before approaching a community, especially if they are both Extension professional and evaluator. Vulnerable communities may hesitate to work with individuals associated with any governmental entity including university personnel. Any perceived connection to the government may lead some individuals to worry about their status or safety. For example, this sentiment may be especially relevant when working in immigrant communities today given the contested climate surrounding illegal immigration and deportation practices. People that have immigrated to the United States from certain countries and/or cultures may be reluctant to share any information with an individual identified as a government employee due to their immigration status (e.g., Armitage, 2008). Despite the challenges, program providers and Extension professionals recognize these individuals can contribute important data as well as being individuals that might benefit from community programs.

Evaluators will want to consider cultural practices and potential barriers, as some communities experience distinct challenges that prevent their involvement in evaluation. Culture is comprised of various beliefs, attitudes, values, morals, and knowledge found within a community. The size of a community is irrelevant to the concept of culture, as the culture develops through the transmission of the previously stated elements. Regardless of the group size, an evaluator should take time to explore the nuances of a community's cultural practices prior to initiating an evaluation (Wehipeihana & McKegg, 2018). Learning a community's culture is time consuming and must extend beyond surface-level interactions. An evaluator will gain greater insight into a culture through firsthand experience. However, gaining access to the target community for an evaluation may not be straightforward.

An evaluator's ability to gain access to a specific population or community will rest on their ability to engage key individuals early in the project-development phase in order to ascertain the cultural norms (Lofland et al., 2006). These key individuals can support the evaluation process by providing suggestions regarding data-collection strategies, identifying important issues within the community, and giving personal endorsements to encourage community members to participate. In any community or population, there is typically an individual or group of people that serve as the gatekeeper(s). Community members trust these key individuals to safeguard the traditions and integrity of their community (Lofland et al., 2006; Wehipeihana & McKegg, 2018). Gatekeepers could be formal and informal. For example, formal gatekeepers are individuals that ultimately determine your access into a community such as clergy, school administrators, or community activists. Upon gaining access to a community, an evaluator may encounter informal gatekeepers while implementing the evaluation project such as head teachers, parents, or team leaders. These informal gatekeepers are the individuals that can rally a group together or completely undermine a project. Initial rapport building begins with the formal gatekeeper; then an evaluator continually develops their rapport within the community with a particular focus on those informal gatekeepers. A multifaceted approach to rapport building will help the evaluator keep the community engaged for an accurate and representative evaluation.

While attempting to gain access to a community, an evaluator is simultaneously developing a rapport with that population. An evaluator's demeanor, attitude, and/or behaviors when interacting with community members may influence their attempt at building a strong rapport (Lofland et al., 2006). A community may be more willing to engage when an evaluator spends time conversing and visiting with individuals before approaching anyone to participate in an evaluation, unlike an evaluator visiting a program for the first time to both recruit participants and complete the evaluation. The amount of time needed to build a positive rapport with a community or population is not standardized. Some people can develop a rapport quickly, while other communities may be more suspicious, resistant, or hesitant to engage with non-community members. Developing a positive rapport with your evaluation participants could lead to increased participation as well as greater opportunity for in-depth information. An evaluator with a positive rapport elicits greater trust and acceptance from a community, especially communities considered to be vulnerable.

One of the best approaches to building a positive rapport is spending time in the community, which could involve participating in programs or volunteering alongside community members. The evaluator's engagement facilitates conversations, demonstrates a personal interest in the community, and displays the motives for initiating an evaluation within the community. This approach is particularly important when working in communities with customs that differ from the dominant population or the ones experienced by the evaluator. For instance, Wehipeihana and McKegg (2018) discovered all significant community discussions occurred in the "Marae" (traditional place of gathering) in the Indigenous community of the Maori. This knowledge was significant since the researchers hoped to gather data through community-member conversations. Similarly, Smalkoski et al. (2016) suggested that community members and Extension partners should cocreate the program or evaluation agenda so that the community remains at the heart of the initiative rather than having the evaluation *done to* the community. The time and effort dedicated to identifying and engaging key individuals as well as spending time in a community before the evaluation can be time consuming. However, these efforts can increase participation levels and collection of higher quality data.

As previously stated, there is no standard amount of time to building rapport, as multiple variables influence the timetable: previous relationships, organizational affiliation, cultural norms, personal demeanor, and attitude. Some evaluators may know their community due to their role as a program provider. This situation may not require additional time to build rapport if the evaluator has a positive reputation within the community. However, the evaluator should still consider the cultural norms and community needs related to potential evaluation engagement. A community may be willing to participate in a program but will hesitate to contribute information on the record for an evaluation regardless of any confidentiality guarantee.

The process used by a local program provider or Extension professional to gain access to a community to provide a program is similar to the process used to conduct a community-level evaluation. An evaluator will entice few individuals by distributing general recruitment flyers, regardless of the language used in the material (Smalkoski et al., 2016). A "generic outreach" effort overlooks nuanced norms or potential challenges present in any culture or community (Smalkoski et al., 2016). Some communities are overly cautious and protective after prior negative experiences with outsiders innocently seeking to "learn" about or "help" their culture (Emerson, 2001).

Evaluation Methodology

There are multiple methods (i.e., needs assessment, program evaluation, and research studies) that Extension and community-level evaluators use to gather data necessary for making decisions related to program design and implementation. For instance, needs assessment and program evaluation are the basis to quality program development and creation. Both a needs assessment and program evaluation employ similar research methods to gather data, such as the utilization of surveys. Typically, evaluators utilize surveys because they are cost effective and efficient.

However, should surveys be used if the target population is uncomfortable speaking the dominate language or has low literacy levels or other disabilities that may limit in-person interactions?

According to the U.S. Census in 2013, approximately 61.6 million individuals, foreign and U.S. born, spoke a language other than English at home and 25.1 million individuals were considered limited English proficient (Zong & Batalova, 2015). Limited English proficiency is defined by the U.S. Census Bureau as anyone above the age of five who reported speaking English less than "very well" (Zong & Batalova, 2015). Use of surveys alone when studying vulnerable populations can easily isolate a specific population simply because they are unable to read or comprehend the language used in the survey. The *Guidelines for Best Practices in Cross-Cultural Surveys* suggests that "studies involving multiple cultures, countries, regions, or languages may benefit from the use of mixed research methods" (Survey Research Center, 2016, p. 7).

A mixed-method approach can increase data quality and validity. Applying both quantitative and qualitative research methods to study the same phenomenon provides validation of data through cross verification while the use of both methods counterbalances the potential limitations of either one individually (Survey Research Center, 2016). Additional studies have found that a mixed methodology was both a cost-effective way to obtain the highest quality data and the best approach to examining issues related to low-income and minority populations (Ngo-Metzger et al., 2004). According to Miller (2010), qualitative research methodologies are becoming increasingly popular among applied health researchers in order to acknowledge the voices of vulnerable populations. Specifically, focus groups have the potential to engage individuals from a more culturally and linguistically diverse background as well as those who have low literacy skills since focus groups are an open format that allow participants to respond verbally in their own words (Halcomb et al., 2007).

Researchers and evaluators may utilize other techniques and technologies (e.g., video conferencing, audio recording, social media, and document-sharing platforms) to engage different populations. For example, participants who may have limited time, ability-level, and/or resources for face-to-face focus groups may have the ability to engage using cell phones and/or video conferencing. Consistent Internet access remains a significant challenge for rural and sparsely populated areas; however, certain technologies allow participants to audio record messages that can be easily shared with the evaluator.

Evaluators should consider these techniques and technologies with populations where limited literacy is a concern. Specifically, the use of audio recording allows individuals with limited literacy to contribute their perspectives to the evaluation, which would be overlooked when only administering a self-report survey. Assistive technologies, such as screen-reader software, captions, voice-recognition software, and many other alternative input devices increase participation for individuals with disabilities. Regardless of perceived ability level, Extension evaluators should adhere to accessibility best practices when developing surveys, recruitment letters, and other supporting resources.

Additional methodologically rigorous and culturally appropriate qualitative research methods that have been adopted with Indigenous populations in Canada include photovoice, Anishinaabe symbol-based reflection, and sharing circles (Wright et al., 2016). Photovoice is a participatory qualitative research method developed by Wang and Burris (1997, p. 370) that enables participants to document through photography the needs, concerns, strengths, and other aspects of their communities, and creates dialogue about those items. Anishinaabe symbol-based reflection "is an arts-based research approach" that "was influenced by Photovoice" (Lavallée, 2009, p. 30). In this method, participants use creative arts to express their lived experiences. Sharing circles allow participants to tell their stories in their own voices, similar to the approach taken with focus groups (Lavallée, 2009, p. 28). These research methods are mutually beneficial for both the researcher and the target population, which is important for cultural appropriateness and sensitivity.

Language and the meaning of words is another important consideration when conducting research with vulnerable populations. For instance, translating a survey from one language to another may be a valuable and important component of data collection. Evaluators must recognize that translating surveys, interviews, focus groups, or any other type of research method can significantly increase the cost of the research study or program evaluation (Lee, Sulaiman-Hill & Thompson, 2014). Yet translation provides an opportunity for participants to use their native language, which could mean the difference between obtaining data or not. There are important recommendations to follow when using translators such as obtaining the required credentials and ensuring the translator has prior experience with professional translators' associations (e.g., American Translators Association and International Association of Professional Translators and Interpreters). The translator's knowledge and skills will affect the data collection and, potentially, the analysis process (Squires, 2008). A high-quality translator has the ability to communicate between languages and describe concepts or words when the translator or participants do not know the actual word or phrase.

Even with quality translation, translators may have difficulty deciphering participants' experiences described in their native language. Cultural and social norms may not easily translate from one language to another (Rubinstein-Avila, 2009). In some cases, it might be appropriate to employ bilingual workers for the duration of the evaluation project. This type of worker may serve as a research assistant who can communicate in the participants' language then communicate in another language with the project researchers (Lee, Sulaiman-Hill & Thompson, 2014). A bilingual research worker may aid with participant recruitment by connecting with individuals who may be more comfortable speaking in their native language, which broadens the voices of vulnerable populations. Moreover, the use of bilingual workers can provide general cultural information, clarify cultural concepts, and even provide context for the information shared and meaning of words and phrases (Lee, Sulaiman-Hill & Thompson, 2014).

A rigorous analysis for qualitative research may engage participants in member checking as a way to safeguard against misinterpretation of data or misrepresentation of perspectives (Merriam, 2009). Member checking validates, verifies, and provides trustworthiness of qualitative results (Doyle, 2007). The process of member checking involves taking the preliminary analysis back to the participant and seeking their feedback on the researcher's interpretation based on their experience (Merriam, 2009). This process can be accomplished by participants checking transcripts, reviewing data analysis such as themes, or reading a summary of data and data analysis (Doyle, 2007). According to Birt et al. (2016) ethical considerations arise during this phase of research, such as do participants fully engage with the research results or do they simply accept what the researcher presented to them? Some populations might be more at risk of not providing adequate feedback based on their relationship with the evaluator (Doyle, 2007) and the evaluator's ethics (Birt et al., 2016).

The write-up of data and its analysis is another important element in program evaluation. During this stage of evaluation, a potential power imbalance might emerge if the evaluator does not maintain clear and transparent communication with the population that was the subject of evaluation (Marshall & Batten, 2004). These populations often have fewer skills for interpreting detailed scholarly manuscripts or lack an understanding of data analysis. A way to continue participant engagement is through coauthoring the reports, which acknowledges community members' involvement in the evaluation (Tondu et al., 2014) and provides a better description of the population's culture and values (Christopher et al., 2008).

The City of Menahga, Minnesota (population 1,306) and University of Minnesota Extension are an example of the use of participatory action research to enable local business success (French & Morse, 2015). Extension field specialists and university-based specialists utilized both surveys and participant interviews for a needs assessment. A fourteen-person research panel consisting of Menahga Task Force members, Extension specialists, and state agency partners reviewed the data and presented action steps in a research report (French & Morse, 2015). Ultimately, the Menahga Task Force determined the action steps and strategies to implement from the needs assessment (French & Morse, 2015). This use of a participatory action research model, or any of the methodological examples discussed in this chapter, highlights how researchers and participants can establish an equal partnership and collaborate to address community needs through evaluation.

Key Elements for Culturally Responsive Evaluation

House and Howe remind us that evaluations sit within sociopolitical contexts. They put forth the theory of deliberative democratic evaluation, which seeks to reduce bias through "inclusion, dialogue, and deliberation" (House & Howe, 2000, p. 5). Inclusion means that the interests of all stakeholders be represented. Discovering stakeholders' "real" interests requires dialogue. Deliberation requires that evaluators and stakeholders engage in thinking based on facts, evidence, and reason. To implement deliberative democratic evaluation, House and Howe offer ten questions that evaluators may answer to ensure all aspects of inclusion, dialogue, and deliberation are front of mind:

1. Whose interests are represented?
2. Are major stakeholders represented?
3. Are any stakeholders excluded?
4. Are there serious power imbalances?
5. Are there procedures for controlling the imbalances?
6. How do people participate in the evaluation?
7. How authentic is their participation?
8. How involved are they?
9. Is there reflective deliberation?
10. How thoughtful and extended is the deliberation? (House & Howe, 2000, pp. 10–11)

While considering these questions, in addition to the questions posed earlier in the chapter, an evaluator seeking to design a culturally responsive evaluation must first engage in self-reflection and self-disclosure. What kind of understanding and cultural knowledge does the evaluator possess to accurately interpret and validate the experience of a given culture being studied (Tillman, 2002)? Additionally, researchers need to crucially examine their own individual values, assumptions, and biases when working with vulnerable populations (Chouinard & Cousins, 2007). Fryer et al. (2011) found that researchers and language assistants engaged in a limited amount of critical self-reflection. Finlay (2002, p. 531) describes reflexivity as the process when researchers "engage in explicit, self-aware analysis of their own role." Through the process of reflexivity, research meanings and findings are "co-constituted" or cocreated between the researcher and the research participants, often allowing for multiple reflective cycles during the research process (Finlay, 2002, p. 535). During this type of self-reflection or reflexivity process, a trusting relationship emerges as all stakeholders contribute their expertise, share decision-making responsibilities, and take ownership of the study design and reporting (Christopher et al., 2008).

Valuable insight for Extension and other community-level programs is lost when vulnerable populations remain unengaged in evaluation. As with all responsible evaluative practices, Extension professionals must adhere to the guiding principles put forth by the American Evaluation Association and university IRBs to engage in the ethical practices of respect for persons, beneficence, and justice; practice self-reflection and reflexivity; use both quantitative and qualitative approaches based on the research questions; consider the social-political environments in which we operate; and respect the cultures and traditions of the research and evaluation participants. Culturally responsive evaluation in Extension allows all populations to lend their voice so that we can more fully understand the individuals, families, and communities we serve and to develop the practices and skills needed in a complex, multicultural world.

REFERENCES

AEA (American Evaluation Association). (2011). American Evaluation Association Statement on Cultural Competence in Evaluation. Approved April 22. Retrieved from https://www.eval.org/About/Competencies-Standards/Cutural-Competence-Statement.

Archibald, T. (2015). "They Just Know": The Epistemological Politics of "Evidence-Based" Non-formal Education. *Evaluation and Program Planning*, 48, 137–148.

Archibald, T., Neubauer, L. C., & Brookfield, S. D. (2018). The Critically Reflective Evaluator: Adult Education's Contributions to Evaluation for Social Justice. *New Directions for Evaluation*, 158, 109–123. https://doi.org/10.1002/ev.

Armitage, J. S. (2008). Persona Not Grata: Dilemmas of Being an Outsider Researching Immigration Reform Activism. *Qualitative Research*, 8(2), 115–177.

Azzam, T., & Levine, B. (2014). Negotiating Truth, Beauty, and Justice: A Politically Responsive Approach. In J. C. Griffith & B. Montrosse-Moorhead (Eds.), Revisiting Truth, Beauty, and Justice: Evaluating with Validity in the 21st Century. Special issue, *New Directions for Evaluation*, 142, 57–70.

Biesta, G. J. J. (2010). Why "What Works" Still Won't Work: From Evidence-Based Education to Value-Based Education. *Studies in Philosophy and Education*, 29(5), 491–503.

Birt, L., Scott. S., Cavers, D., Campbell, C., & Walter, F. (2016). Member Checking: A Tool to Enhance Trustworthiness or Merely a Nod to Validation? *Qualitative Health Research*, 26(13), 1802–1811.

Bledsoe, K. L. (2014). Truth, Beauty, and Justice: Conceptualizing House's Framework for Evaluation in Community-Based Settings. In J. C. Griffith & B. Montrosse-Moorhead (Eds.), *Revisiting Truth, Beauty, and Justice: Evaluating with Validity in the 21st Century*. Special issue, *New Directions for Evaluation*, 142, 71–82.

CDC (U.S. Centers for Disease Control & Prevention). (2015). U.S. Public Health Service Syphilis Study at Tuskegee. https://www.cdc.gov/tuskegee/timeline.htm.

Chouinard, J., & Cousins, J. B. (2007). Culturally Competent Evaluation for Aboriginal Communities: A Review of the Empirical Literature. *Journal of MultiDisciplinary Evaluation*, 4(8), 40–57.

Christopher, S., Watts, V., McCormick, A., & Young, S. (2008). Building and Maintaining Trust in the Community-Based Participatory Research Partnership. *American Journal of Public Health*, 98(8), 1398–1406.

Coalition for Evidence-based Policy (2003). *Identifying and Implementing Educational Practices Supported by Rigorous Evidence: A User-Friendly Guide*. Washington, DC: U.S. Dept. of Education Institute of Education Sciences National Center for Education Evaluation and Regional Assistance.

Donaldson, S. I., Christie, C. A., & Mark, M. M. (Eds.) (2009). What Counts as Credible Evidence in Applied Research and Evaluation Practice? Los Angeles: Sage.

Doyle, S. (2007). Member Check with Other Women: A Framework for Negotiating Meaning. *Health Care for Women International*, 28, 8888–8908.

Emerson, R. M. (2001). *Contemporary Field Research Perspectives and Formulations* (2nd ed.). Long Grove, IL: Waveland Press.

Finlay, L. (2002). "Outing" the Researcher: The Provenance, Process and Practice of Reflexivity. *Qualitative Health Research*, 12(4), 531–545.

Fitzpatrick, J. L. (2012). An Introduction to Context and Its Role in Evaluation Practice. In D. J. Rog, J. L. Fitzpatrick, & R. F. Conner (Eds), Context: A Framework for Its Influence on Evaluation Practice. Special issue, *New Directions in Evaluation*, 135, 7–24.

French, C., & Morse, G. (2015). Extension Stakeholder Engagement: An Exploration of Two Cases Exemplifying 21st Century Adaptions. *Journal of Human Sciences and Extension*, 3(2), 108–131.

Fryer, C., Mackintosh, S., Stanley, M., & Crichton, J. (2011). Qualitative Studies Using In-depth Interviews with Older People from Multiple Language Groups: Methodological Systematic Review. *Journal of Advanced Nursing*, 68(1), 22–35. https://doi.org/10.1111/j.1365-2648.2011.05719.x.

Greene, J. C. (2009). Evidence as "Proof" and Evidence as "Inkling." In S. I. Donaldson, C. A. Christie, & M. M. Mark (Eds.), *What Counts as Credible Evidence in Applied Research and Evaluation Practice?* (pp. 153–167). Los Angeles: Sage.

Halcomb, E. J., Gholizadeh, L., DiGiacomo, M., Phillips, J., & Davidson, P. M. (2007). Literature Review: Considerations in Undertaking Focus Group Research with Culturally and Linguistically Diverse Groups. *Journal of Clinical Nursing*, 16, 1000–1011. https://doi.org/10.1111/j.1365-2702.2006.01760.x.

HHS (U.S. Department of Health and Human Services). (1979). The Belmont Report: Ethical Principles and Guidelines for the Protection of Human Subjects of Research. https://www.hhs.gov/ohrp/regulations-and-policy/belmont-report/read-the-belmont-report/index.html.

House, E. R., & Howe, K. R. (2000). Deliberative Democratic Evaluation. *New Directions for Evaluation*, 85, 3–12.

Johns Hopkins Medicine. (n.d.). *The Legacy of Henrietta Lacks*. https://www.hopkinsmedicine.org/henriettalacks/upholding-the-highest-bioethical-standards.html.

Lavallée, L. F. (2009). Practical Application of an Indigenous Research Framework and Two Qualitative Indigenous Research Methods: Sharing Circles and Anishnaabe Symbol-Based Reflection. *International Journal of Qualitative Methods*, 21–40. https://doi.org/10.1177/160940690900800103.

Lee, S. K., Sulaiman-Hill, C. R., & Thompson, S. C. (2014). Overcoming Language Barriers in Community-Based Research with Refugee and Migrant Populations: Options for Using Bilingual Workers. *BMC International Health and Human Rights*, 14(11). https://doi.org/10.1186/1472-698X-14-11.

Lofland, J., Snow, D., Anderson, L., & Lofland, L. H. (2006). *Analyzing Social Settings: A Guide to Qualitative Observation and Analysis* (4th ed.). Belmont, CA: Wadsworth.

Madison, A. (2007). *New Directions for Evaluation* Coverage of Cultural Issues and Issues of Significance to Underrepresented Groups. *New Directions for Evaluation*, 114, 107–114. https://doi.org/10.1002/ev.227.

Marshall, A., & Batten, S. (2004). Researching across Cultures: Issues of Ethics and Power. *Forum: Qualitative Social Research*, 5(3).

McBride, D. F. (2011). Sociocultural Theory: Providing More Structure to Culturally Responsive Evaluation. *New Directions for Evaluation*, 131, 7–13. https://doi.org/10.1002/ev.371.

Merriam, S. B. (2009). *Qualitative Research: A Guide to Design and Implementation*. San Francisco, CA: Jossey-Bass.

Miller, W. R. (2010). Qualitative Research Findings as Evidence: Utility in Nursing Practice. *Clinical Nurse Specialist*, 24(4), 191–193.

Ngo-Metzer, Q., Kaplan, S. H., Sorkin, D. H., Clarridge, B. R., & Phillips, R. S. (2004). Surveying Minorities with Limited English Proficiency: Does Data Collection Method Affect Data Quality among Asian Americans? *Medical Care*, 42(9), 893–900.

Rubinstein-Avila, E. (2009). Reflecting on the Challenge of Conducting Research across National and Linguistic Borders: Lesson from the Field. *Journal of Language and Literacy Education*, 5(1), 1–8, https://files.eric.ed.gov/fulltext/EJ1068181.pdf.

SenGupta, S., Hopson, R., & Thompson-Robinson, M. (2004). Cultural Competence in Evaluation: An Overview. *New Directions for Evaluation*, 102, 5–19. https://doi.org/10.1002/ev.112.

Skloot, R. (2010). *The Immortal Life of Henrietta Lacks*. New York: Crown Publishers.

Smalkoski, K., Axtell, S., Zimmer, J., & Noor, I. (2016). One Size Does Not Fit All: Effective Community-Engaged Outreach Practices with Immigrant Communities. *Journal of Extension*, 54(4), 4COM1.

Smith, L. T. (2012). Decolonizing Methodologies: Research and Indigenous People (2nd ed). London: Zed Books.

Squires, A. (2008). Methodological Challenges in Cross-Language Qualitative Research: A Research Review. *International Journal of Nursing Studies*, 46, 277–287.

Survey Research Center. (2016). *Guidelines for Best Practice in Cross-Cultural Surveys*. Ann Arbor, MI: Survey Research Center, Institute for Social Research, University of Michigan. http://www.ccsg.isr.umich.edu/.

Symonette, H., Miller, R. L., & Barela, E. (2020). Power, Privilege, and Competence: Using the 2018 AEA Evaluator Competencies to Shape Socially Just Evaluation Practices. *New Directions in Evaluation*, 168, 117–132. https://doi.org/10.1002/ev.20433.

Tillman, L. C. (2002). Culturally Sensitive Research Approaches: An African-American Perspective. *Educational Researcher*, 31(9), 3–12.

Tondu, J. M. E., Balasubramaniam, L., Gantner, C. N., Knopp, J. A., Provencher, J. F., Wong, P. B. Y., & Simmons, D. (2014). Working with Northern Communities to Build Collaborative Research Partnerships: Perspectives from Early Career Researchers. *InforNorth*, 67(3), 419–429.

Vo, A. T., & Archibald, T. (2018). New Directions for Evaluative Thinking. In Evaluative Thinking, special issue, *New Directions for Evaluation*, 158, 139–147. https://doi.org/10.1002/ev.20317.

Wang, C., & Burris, M. A. (1997). Photovoice: Concept, Methodology, and Use for Participatory Needs Assessment. *Health Education & Behavior*, 24(3), 369–387.

Wehipeihana, N., & McKegg, K. (2018). Values and Culture in Evaluative Thinking: Insights from Aotearoa New Zealand. *New Directions for Evaluation*, 158, 93–107. https://doi.org/10.1002/ev.

Wright, A., Wahoush, O., Ballantyne, M., Gabel, G., & Jack, S. (2016). Qualitative Health Research Involving Indigenous Peoples: Culturally Appropriate Data Collection Methods. *The Qualitative Report*, 21(2), 2230–2245.

Zong, J., & Batalova, J. (2015). The Limited English Proficient Population in the United States. Migration Policy Institute, Washington, DC. https://www.migrationpolicy.org/article/limited-english-proficient-population-united-states.

PART 2

Grassroots Engagement in Practice

The USDA and Land-Grant Extension: A Legacy of Continuing Inequities

Robert Zabawa and Lindsey Lunsford

Since their creation, within months of each other in 1862, the U.S. Department of Agriculture (USDA) and the land-grant system (LGS) have been inextricably bound together. In the first case, because the majority of the U.S. population was both rural and involved in agriculture, President Abraham Lincoln promoted and signed legislation to create an Agriculture Department in May 1862. In his last State of the Union address, Lincoln referred to it as the "people's Department":

> The Agricultural Department . . . is rapidly commending itself to the great and vital interest it was created to advance. It is peculiarly the people's Department, in which they feel more directly concerned than in any other. I commend it to the continued attention and fostering care of Congress. (Lincoln, 1864)

At the same time, higher education, traditionally the realm for the upper classes and elites, was beyond the grasp of the common person, including farmers and ranchers. To address this issue, legislation proposed by Congressman Justin Morrill of Vermont was passed in July 1862 to establish the first land-grant colleges.

As time progressed, however, it became apparent that "the people's Department" and the land-grant system of education were not open for all people in the Southern United States starting in the post-Reconstruction era. Both ideals of agriculture and education were beset by inequities based on race, and the very foundation of the LGS was rooted in legislative inequity, starting with section 1(2) of the Morrill Act of 1890 that allowed separate but equal land-grant institutions to coexist in inequality (Land Grants, 2019b, section 1[2]).

Legislative inequity continued with the Hatch Act of 1887 for land-grant research and culminated with the Smith-Lever Act of 1914 for land-grant Extension that allowed the states to decide how research and Extension funds were allocated. Specifically for Extension, Section 1 of the Smith-Lever Act prescribed the allocation of funding as follows:

> in any State, Territory, or possession in which two or more such colleges have been or hereafter may be established, the appropriations [for agricultural Extension work] hereinafter made to such State, Territory, or possession shall be administered by such college or colleges as the legislature of such State, Territory, or possession may direct. (Land Grants, 2019d)

It is because the states had the power to direct the administration of federal funds to the land-grant agricultural experiment stations and the Cooperative Extension Services that the 1862 institutions held virtual control of all research and Extension funds and that the 1890s were shut out from any support until the 1960s (Mayberry, 1989). And permanent funding for 1890s did not occur until the 1970s and through subsequent farm bills in sections 1444 (for Extension) and 1445 (for research) of the Evans-Allen Act (Rickenbach et al., 2013).

A similar critical issue to control of federal funding is the state match requirement codified under section 1449 of the 1977 Farm Bill. Recently, under its final rule for matching funds for Extension and research, the USDA stipulates that (1) federal funds must be matched by the states "equal and not less than 100 percent" of federal funds; (2) the secretary of agriculture may waive the match above the "50 percent level" if the "Secretary determines that the State will be unlikely to satisfy the matching requirement"; and (3) reasons for failure to provide the match include "impacts from natural disasters, flood, fire, tornado, hurricane, or drought; state and/or institution facing a financial crisis; or lack of matching funds after demonstration of good faith efforts to obtain funds" (USDA, 2018, p. 21849). Additionally, under its call for Extension Applications for fiscal year (FY) 2020, the National Institute of Food and Agriculture (NIFA) provided that they "may consider and approve matching waiver requests above the 50 percent level" (USDA/NIFA, 2019).

The final rule allows the states ample opportunities to declare either a financial crisis or lack of funds despite good faith efforts to not match federal appropriations. The impact of the lack of state matching funds was highlighted by Lee and Keys in their 2013 Policy Brief *Land-Grant but Unequal* for the Association of Public and Land-grant Universities. They examined state matching funds over a three-year period, 2010–2012. In summary, the report highlighted that (1) ten of the eighteen 1890 land-grant institutions did not receive the mandated 100 percent state match for Extension, ranging from Lincoln University at 47 percent to Virginia State University at 68 percent, and this resulted in an annual loss of $10.6 million to these ten institutions; (2) for research, the range in state match was 42 percent for Florida A&M and South Carolina State Universities to 95 percent for Tuskegee University, and this resulted in an annual loss of $8.3 million for these ten institutions; and (3) the combined annual loss to the ten 1890 institutions due to the lack of state matching funds was $18.9 million (see table 1).

By FY 2020, there was relatively little change in mandated state support for the 1890 land-grant institutions. Six of the eighteen 1890 land-grant institutions did not receive the mandated 100 percent state match for Extension, ranging from Alcorn University in Mississippi and Prairie View A&M University in Texas at 50 percent to North Carolina A&T State University at 92 percent, and this resulted in an annual loss of $7.8 million to these six institutions. For research, the range in state match was again from 50 percent with Langston University in Oklahoma joining Alcorn and North Carolina A&T to 83 percent for Tuskegee University, and this resulted in an annual loss of $10.8 million for these seven institutions. The combined annual loss to these 1890

Table 1. State Match for 1890 Research and Extension: FY 2010–2012 (averaged) and FY 2020

CATEGORY	TOTAL TO STATES AND 100% MATCH REQUIRED	TOTAL WAIVER REQUESTED	STATE ACTUAL MATCH	STATE ACTUAL MATCH
EXTENSION				
APLU	$40,915,407	$10,609,639	$30,305,768	74%
2020	$54,720,000	$7,839,550	$46,880,450	86%
RESEARCH				
APLU	$40,702,402	$8,266,094	$32,436,308	80%
2020	$62,910,320	$10,806,133	$52,104,187	83%
COMBINED				
APLU	$81,617,809	$18,875,733		77%
2020	$117,630,320	$98,984,637		84%

Sources: For FY 2010–2012, Lee and Keys (2013); for FY 2016, USDA/NIFA (2020).

institutions in FY 2020 due to the lack of state matching funds for Extension and research was $18.6 million (see table 1).

In summary, the creation of the LGS was based on inequity. Starting with the Morrill Acts of 1862 and 1890, "separate but equal" land-grant institutions were created based on race. The other two missions of the LGS, established by the Hatch Act of 1887 for research and the Smith-Lever Act of 1914 for Extension, allowed the respective states to decide where federal dollars were sent—through the 1862 institutions. And finally, the states were given every opportunity to avoid appropriating the 100 percent match to federal dollars for research and Extension to the 1890 institutions.

Despite the fact that the 1890 land-grant institutions have suffered, and continue to suffer, the lack of federally mandated financial support, they have proven over the years to be the backbone of outreach for both their region and the nation. Three case studies are presented to show how 1890 institutions fulfilled their land-grant mission in times of national crisis. In the first case, 1890 Extension was needed to increase food production from African American farmers during World War II. In the second case the 1890s were a conduit to provide outreach and technical assistance programs to socially disadvantaged farmers and ranchers, long neglected by the USDA and LGS. Finally, the third case serves as an example of how 1890 Extension can be used to support local food systems.

Case 1: World War II, USDA, and the African American Farmer

Despite the fact that almost total control of the land-grant aspects of the 1890 institutions was under the 1862 institutions as directed by the states, in many cases these same institutions were called upon to provide direct support for national emergency needs. In 1942, the agricultural sector was critical because the United States found itself having to supply food and fiber needs not only for the domestic consumer, but also for allies in Europe and Asia (civilian and military) as well as for the U.S. military, and African American farmers were called upon to do their share for the national welfare. Toward this end, in a letter dated February 25, 1942, Secretary of Agriculture Claude R. Wickard requested Tuskegee Institute president Frederick Douglass Patterson

> to accept a position in my immediate office as special assistant to the Secretary. . . . In the position which I am asking you to accept, you will, I believe, be able to make a fine contribution to the general welfare of Negro farmers. And in our current war effort, your work of helping to promote the full participation of Negro farmers in the food-for-freedom campaign will mean for them a better diet and better health. Now, as always, I feel that Negroes, both on the farm and in our cities and towns, are ready to do their full share. I have no doubt of their loyalty and devotion. But, like all our people, their energies must be implemented and directed so that they may do the things which count most toward final victory. (Tuskegee University Archives)

Secretary Wickard also asked Claude A. Barnett, director of the Associated Negro Press, to work closely with Patterson in this endeavor. Barnett was an interesting choice for two reasons: as a newsman, their activities would be well publicized, and also Barnett was a member of the Tuskegee Board of Trustees and a champion of the young Patterson when he was selected as Tuskegee's third leader. Both Patterson and Barnett soon accepted Wickard's offer and set off on a tour of the South in the spring of 1942. They interviewed land-grant officials from both 1862 and 1890 institutions, Cooperative Extension personnel, USDA agency personnel, and farmers. Their mission was to see what the biggest issues were pertaining to meeting ever-increasing USDA production goals.

In a letter dated June 26, 1942, Patterson and Barnett submitted a "preliminary" report of their activities and findings. For brevity's sake here, the series of recommendations can be distilled into two major points. If increased production by Negro farmers was desired, then they would need (1) better instruction and Extension and (2) more Negro Extension and demonstration agents to instruct them. Secretary Wickard's reply by letter four days later, on June 30, was not encouraging: "As you know, we have not met with too much success in getting additional appropriations for Extension work." Patterson also continued to work with Barnett on the issue of improving Negro farm productivity, which had become a funding issue. The bottom line was that current spending on the Negro Extension force was a little over $1 million; a working budget for what Patterson and Barnett proposed was a little over an additional $1.5 million to the current appropriation.

Given the previous concern by Secretary Wickard about the inability to get additional funding for Extension, he had Reuben Brigham, an assistant secretary of agriculture, join the efforts of Patterson and Barnett who also included the presidents of the Negro land-grant colleges. Several ideas came out of the meetings held in the fall of 1942. First, throughout all of their travels, from the Southern states to the national Extension service offices in Washington, DC, Patterson and Barnett were met with positive feedback for their advocacy of increasing the number of Negro agents. Second, as it stood, no further funding was going to be made available to hire new Negro agents. And third, assurances were given that there was no proposal on the table to create a separate Negro Extension service. This last stipulation was considered essential if any support from Southern legislators was to be possible.

In summary, the results of Patterson's efforts were mixed by the end of 1942. Patterson and Barnett were fulfilling their mandate with a plan, including a budget, to increase much-needed agricultural production for the war effort, both domestically and overseas. Additionally, there appeared to be agreement on all sides to the plan itself, but no political will to push it through Congress by the secretary's office.

However, the question remained about what happened to Patterson's proposal for more Negro Extension agents. Interestingly, in an effort to meet with President Franklin D. Roosevelt, Patterson met and later began a correspondence with First Lady Eleanor Roosevelt concerning his quest for support for more Negro agents. This occurred while continued negotiations with the USDA appeared to be agonizingly slow on the issue of support for more agents. This culminated in the winter of 1943 when Secretary Wickard had praise for the Negro farmers' production efforts, but still no funds for more agents to assist them.

But 1944 was an auspicious year. The tide turned in the war effort, which was also problematic, since it was always a pressure point for obtaining emergency funds to support the Patterson proposal. In January, an emergency allocation of $2 million was made available to State Extension Services by the War Food Administration to stimulate increased production and conservation of food and fiber. From this fund, 272 Negro emergency agricultural aids were added to the State Extension Service staffs throughout the South to help push wartime food production and conservation in rural and urban areas (see Schor, 1986).

In conclusion, there was no real change in the status quo. Any success by Patterson (and others) is with a small "s." The United States needed the help of all of its citizens to win the war, and its African American citizens were, as they always had been, willing and able to contribute. However, they needed the straw of equitable resources to make the bricks, or in this case, added food, for the nation to be full participating members of society. A turn to such equity was not to occur for another generation.

Case 2: The 2501 Program and Outreach to the African American Farmer

The Outreach and Assistance for Socially Disadvantaged and Veteran Farmers and Ranchers (OASDVFR) Program is also called the 2501 Program because of its placement in the farm bill, under Title XXV, section 2501 of the Food, Agriculture, Conservation, and Trade Act of 1990 (Pub.L. 101-624). Its true beginning, however, was in 1981, when President Ronald Reagan signed executive order 12320 in support of "Historically Black Colleges and Universities," which in turn led to the creation of the Small Farmers Outreach Training and Technical Assistance Program of 1983. Support for this program came from two major sources, the 1890 community and community-based organizations (CBOs) from the South, including, for example, the Federation of Southern Cooperatives.

The initial goals of this program were to assist socially disadvantaged farmers and ranchers with farm and financial management, access to USDA programs, and a linkage with USDA offices. Assistance was provided by institutions of higher education (e.g., land-grant universities) and CBOs that served socially disadvantaged farmers and ranchers. Support through grants for this program was through the Farmers Home Administration (FmHA). The first grant was awarded to

North Carolina A&T State University in 1983, and then later expanded in 1984–1987 to Tuskegee University in Alabama, Fort Valley State University and the Federation of Southern Cooperatives in Georgia, and New Mexico State University in New Mexico. Later legislative provisions in support of assistance included the Food Security Act of 1985 (Pub.L. 99-198, also known as the 1985 Farm Bill), and the Agricultural Credit Act of 1987 (Pub.L. 100-233).

By 1994, activities by FmHA relative to the 2501 program were taken over by the Consolidated Farm Services Agency (CFSA), and an interim rule was published to establish procedures for grants to assist farmers through the program. According to the interim rule, the objective of 2501 was to "reverse . . . the decline of socially disadvantaged farmers and ranchers across the United States." The outcomes of the program were to have farmers own and operate their own farms, to have them participate in agricultural programs, and to have them become members of the agricultural community. These outcomes were to be achieved through "outreach training and technical assistance in farm and ranch management, record keeping, marketing techniques, and testing innovative solutions." Two additional major points in the interim rule included the length of the project period (five years) and the project administrators.

Like his predecessor before him, President William J. Clinton signed an executive order (12876) in support of "Historically Black Colleges and Universities" (HBCUs). One of the objectives of this order was that HBCUs (i.e., 1890 institutions) would become "a major beneficiary of this funding [for 2501]." It is significant, however, that given the support of two presidents, Reagan and Clinton, for HBCUs through executive orders, and given the vehicle for this support through HBCU/1890 outreach, and given a clientele in need of support, African American farmers, that the record of the 2501 program is mixed.

The director of the USDA Office of Partnerships and Public Engagement shares on the 2501 webpage:

> All farmers and ranchers deserve equal access to USDA programs and services. 2501 grants go a long way in fulfilling our mission to reach historically underserved communities and ensure their equitable participation in our programs. (USDA, 2019b)

Yet, despite the successes and the support from the community level to USDA offices, the 2501 program has had challenges that keep it from achieving major structural success in the field. Three major challenges are funding at the authorized level, consistent management, and continuity of programs.

Funding

Ever since the 2501 program started releasing grants in 1994, appropriated funds have never approached authorized funds. From 1994 to 2001, the Farm Bill authorized $10 million annually for 2501, yet only an average of $4.65 million was appropriated for twenty-seven projects. Starting in 2002 through 2008, 2501 was authorized for $25 million per year, yet only an average of $4.8 million was appropriated. From 2009 to 2012, authorization remained at $25 million, but again appropriations did not reach that level. From 2014 through 2019, annual authorized funding was set at $20 million and annual appropriations were set at $10 million, with $9 million set aside for 1890 institutions, 1994 or Native American land-grant institutions, and Hispanic-serving institutions and CBOs, and $1 million set aside for other institutions of higher education, including 1862 institutions.

Consistent Management

Since its creation as a pilot program in 1983, the 2501 program has been managed by nine USDA offices and agencies: Farmers Home Administration (FmHA), Consolidated Farm Service Agency (CFSA), Farm Service Agency (FSA), Natural Resources Conservation Service (NRCS), Departmental Administration, Office of Outreach (DA), Cooperative State Research, Education and Extension Service (CSREES), National Institute of Food and Agriculture (NIFA), Office of Advocacy and Outreach (OAO), and Office of Partnerships and Public Engagement (OPPE). One major issue that has evolved is that, until recently, leadership of programs such as 2501 have been political appointees who have little expectation of long-term tenure, have little experience in outreach for socially disadvantaged clientele, and therefore have little reason to fight for the program itself (also see Hargrove, 2002, p. 123; Hargrove and Jones, 2004).

Continuity of Programs

Related to the authorization/appropriation gap is the 2501 project period. As stated in the interim rule of 1994, 2501 grants were set for a five-year program period. Yet while grant proposals may have been multiyear, actual budgets were from year to year. Starting with the 2018 Farm Bill, multiyear projects were again instituted. Significantly, starting in the mid-2010s, appropriations were cut so deeply that budgets for the annual grants were cut in half from a maximum of $400,000 to $200,000.

Related to financial support and management is the ability of the individual projects to maintain continuity over the long term. That is, if support is not at authorized levels, if management is in a state of flux, and if commitments are on a year-to-year timetable, then projects lose personnel and have a hard time developing rapport with the people they are charged to serve. According to Hargrove, unstable funding negatively affected the six projects in her research through the inability to maintain staff, causing temporary shutdowns, releasing of employees, loss of office space, the inability to provide services to clients, and creating adverse relationships with farmers, the surrounding community, and other USDA agencies (Hargrove 2002, p. 121).

Another factor to consider is the original intent of the program to support HBCUs and African American farmers in the South. From 1994 to 2000 there were twenty-seven 2501 projects at seventeen 1890 institutions, two at 1862 institutions, four at 1994 institutions, three at CBOs, and one at another institution of higher education. By the last round of approved projects in 2019, there were thirty-three 2501 projects at five 1890 institutions, one at an 1862 institution, two at 1994 institutions, twenty-three at CBOs, and two at other institutions of higher education (see table 2). A simple chi-square analysis of this shift from 1890-led to CBO-led projects is a highly significant shift from the initial policy.

Table 2. Distribution of 2501 Projects by Institution Award

1994–2000			2019		
ORGANIZATION	#	%	ORGANIZATION	#	%
1890	17	63.0	1890	5	15.2
1862	2	7.4	1862	1	3.0
1994	4	14.8	1994	2	6.1
CBO	3	11.1	CBO	23	69.7
Other	1	3.7	Other	2	6.1
TOTAL	27	100.0	TOTAL	33	100.1

Notes: $\chi^2 = 22.892$; d.f. = 4; P = 0.0001.
Sources: For the 1994–2000 data, USDA (2001). For the 2019 data, USDA (2019a).

Table 3. Distribution of 2501 Projects by Region Award

1994–2000			2019		
Region	#	%	Region	#	%
South	19	70.4	South	13	39.4
Northeast	3	11.1	Northeast	6	18.2
North Central	5	18.5	North Central	5	15.2
West	0	0.0	West	9	27.3
TOTAL	27	100.0	TOTAL	33	100.1

Notes: $\chi^2 = 10.631$; d.f. = 3; P = 0.01.
Sources: For the 1994–2000 data, USDA (2001). For the 2019 data, USDA (2019a).

Similarly, from 1994 to 2000, of the twenty-seven 2501 projects, nineteen were from the South, three were from the Northeast, and five were from the North Central region of the United States. By the last round of approved projects in 2019, there were thirteen 2501 projects from the South, six from the Northeast, five from the North Central region, and nine from the West (see table 3). This change from the South to other regions, particularly the West, is a significant shift from the initial policy.

Recommendations

To address this history of inequity, two potential solutions are proposed, one financial, one structural. First, financial—reparative payments. If one were to take an approximate annual waiver request of $18 million as highlighted in table 1, and multiply this amount over the forty-three years since federal support for 1890 research and Extension was legislated in the 1977 Farm Bill and the Evans-Allen Act, the total would come to $774 million. Of course, this is a lower limit figure given the fact that the 1890 institutions did not receive the mandated match previous to 1977. The original Hatch Act stated that in those states or territories where two colleges of agriculture existed "appropriations . . . shall be equally divided between such colleges, unless the legislatures of such States or Territories shall otherwise direct." By 1914, the Smith-Lever Act did not include the "equally divided" provision. This should be followed with equity in state matching support.

For example, if the 1862 institution receives a 100 percent match, then the 1890 institution will receive a similar match. Any shortfall in the 1890 institution would result in the proportionate shortfall with the 1862 institution.

Similarly, over the history of the 2501 program, appropriated funds have not matched authorized funds. Indeed, the difference between appropriation and authorization is over $300 million. These are funds that could have been directed at farms, ranches, and surrounding communities that have not been reached. This also means that those institutions and organizations that have a history of serving these clients and communities, by necessity, must compete with each other for scarce resources. For example, the shift in organizational priority to nongovernmental organizations and CBOs, as well as the shift toward a more regional diversity, was mandated in the 2018 Farm Bill (Title XII, subtitle C, section 12301, subparagraphs F and G). This does not mean that the 1890s do not require program assistance to address farm program inequities, nor does it mean that the historic issues that predominated in the South have been resolved.

What this means is that while the number of other agricultural groups added to the program has grown, the financial support for the program itself has not. Indeed, the current budget would not support a 2501 program at all of the 1890 institutions, let alone the other project groups. This also sets up a competition between groups who can justifiably prove the necessity of this program for their clientele. In this case, half a loaf is not better than no loaf when it comes to planning, hiring, and establishing rapport and trust with those in need. It should be noted that the 2018 Farm Bill has brought about significant changes to the 2501 program, including multiyear projects, and authorized funding of $50 million to be divided with the Beginning Farmer and Rancher Program, with progressing mandatory funding to reach this level by FY2023.

The second solution involves the structure of the LGS, and the 1890s specifically. In his 1998 book *Disparity: An Analysis of the Historical, Political, and Funding Factors at the State Level Affecting Black Academic Agriculture*, R. Grant Seals examined how the lack of support to the 1890s resulted in a lack of graduate-level programs and the subsequent lack of a cadre of graduate-degreed African Americans. One of his proposals, which is still relevant today, is to create support for an agricultural "pipeline" in higher education for African American agriculturalists (Seals, 1998, p. 88). Such a program would include graduate training through the PhD (Seals, 1998, p. 93). This proposal could be expanded to where each 1890 institution would select a specialty or specialties, and students would receive federal support for both undergraduate and graduate degrees via scholarships, and specialty-based faculty would receive endowed professorships. This would establish a cadre of researchers necessary for continued needs in teaching, research, and Extension.

Financial commitment and continuity also need to be established for the 2501 program. Priority needs to be set for the establishment of 2501 programs at each 1890 institution and those CBOs who have focused their programs in the rural South. Pennick (n.d.) has proposed a similar recommendation of increased 2501 support with a focus on 1890s and CBOs.

In conclusion, the price of inequity is steep. On the one hand, the *Pigford v. Glickman* settlement was over $1 billion, yet it covered a relatively small period of time (1981–1996). Subsequent class actions, *Keepseagle v. Vilsack* for Native American bias, *Garcia v. Vilsack* for Latinx/Hispanic American bias, and *Love v. Vilsack* for gender bias, have added billions more to the price of inequity. On the other hand, inequity in the LGS was codified with the Second Morrill Act (1890), administered through the Hatch Act (1887) and the Smith-Lever Act (1914), and includes current programs such as 2501 (1990). As Seals (1986, p. 226) comments: "How could a developmental tool (the Land Grant System) specifically designed for the betterment of the 'industrial classes' be denied to those persons most in need?" In the end, the question may be not only how much is owed, but also *how long* until the structure that created inequity is changed and *how* will this change be manifested so that all may benefit and the ideal of Lincoln's "peoples Department" come to fruition.

Integrating Ethnic and Cultural Food into the System: The Black Belt

The first two sections of this chapter highlight financial and structural costs of inequity within the LGS. This section follows by examining the cultural and ethnic costs of inequity within the LGS. This section thus explores the role of Extension services in ensuring culturally and ethnically appropriate food systems. The U.S. Black Belt serves as the primary region of focus as we explore food as a vehicle.

Due to the vestiges of slavery, the breakdown of the antebellum South, the New Jim Crow, and the resurgence of racial violence, the Black Belt South suffers problems of persistent health disparity, social injustices, lack of resources, and racial divide (Wimberley & Morris, 1997). The Black Belt represents a crescent-shaped cluster of 603 counties spread throughout thirteen states in the southeastern United States (Davis, 2000).Following the passage of the Indian Removal Act of 1830 that forcibly removed Native Americans living in the area, the Black Belt emerged as the center of a rapidly expanding plantation economy (Tullos, 2004). The region's name "Black Belt" once referred to the unusually fertile soils of the area. However, as plantation agriculture grew the name began to refer to the increasing presence of enslaved African Americans brought to work the lands. Booker T. Washington explains the term in his seminal text *Up from Slavery*,

> I have often been asked to define the term "Black Belt." So far as I can learn, the term was first used to designate a part of the country which was distinguished by the colour of the soil. The part of the country possessing this thick, dark, and naturally rich soil was, of course, the part of the South where the slaves were most profitable, and consequently they were taken there in the largest numbers. Later, and especially since the war, the term seems to be used wholly in a political sense—that is, to designate the counties where the black people outnumber the white. (Washington, 1901, p. 72)

Contemporarily, Black Belt counties retain concentrations of African Americans exceeding 24 percent of the population or twice the national average (Wimberley, Morris & Bachtel, 1991 as cited in Davis, 2000).

As documented in the first section of this chapter, the USDA and the LGS have been complicit through the lack of support for programs and educational institutions (i.e., the 1890s) created to assist African Americans and the rural poor post-1862. Rather than put the onus on the large structural issues that hinder the quality of life for African Americans living in the Black Belt South, many studies link poor African American health outcomes with individual behavior and traditional African American food customs such as consuming foods high in salt, fat, and cholesterol and fried foods. Other research suggests, however, that this correlation is invalid and unjust to African American communities and the lexicon that is African American cultural food ways (Whitehead, 1992).

The history of Southern ethnic and cultural foods is inextricable from the history of racism in America (Deetz, 2018). So what then is the role of Cooperative Extension in the preservation and integration of ethnic and cultural food into the current system? Any attempt at answering this question requires a journey to the Black Belt, the cradle of Cooperative Extension itself. The Black Belt, the Alabama Black Belt specifically, unequivocally birthed Cooperative Extension, beginning in 1906 with Thomas Monroe Campbell, an African American man, as the first federally appointed Extension agent in the United States at the now Tuskegee University (Mayberry, 1989). In 1906, Seaman A. Knapp, special agent in charge of Farmers' Cooperative demonstration work for the USDA, visited George Washington Carver at Tuskegee (Mayberry, 1989). Knapp engaged Dr. Carver and his staff around the idea of beginning a "co-operative demonstration program for the Negro farmers of the South" (Mayberry, 1989, p. 71). Knapp, using federal funds, then agreed to share expenses to employ a man to conduct demonstration work in Macon County, Alabama, and surrounding Black Belt counties. Recent Tuskegee graduate Thomas M. Campbell secured this position, and in doing so cemented his place in history as the first Cooperative Extension agent. As Mayberry chronicles:

> It is noted, that the term "Cooperative Extension Work" did not come into usage until after the passage of the Smith-Lever Act in 1914. In this case, the definition of Cooperative Extension work is applicable to developments at Tuskegee University on November 12, 1906. Thus, the first cooperative Extension program in the United States emerges at Tuskegee University, and T.M. Campbell became the first Cooperative Extension Agent. (Mayberry, 1989, p. 71)

It is in the Black Belt where you can find an abundance of African Americans, many of whom are direct descendants of the enslaved peoples whose actions, agrarian knowledge, pain, and resilience birthed the modern food system we now know. In addition to a higher than national average of African American residents, the Black Belt is home to persistent poverty, unemployment, underemployment, limited education, lagging infrastructure, crippling dependence on public assistance programs, and abysmal health disparities (Wimberley et al, 2014).

A precarious balance exists within African American foodways and food systems that teeters between pride and pain, erasure and exposure, decadence and despair. *Merriam-Webster* defines foodways as "the eating habits and culinary practices of a people, region, or historical period." This is the definition we will follow in this discussion. It is in these spaces of vast contradiction that Cooperative Extension must find its mantle in the reckoning of it all. If Cooperative Extension wants to understand current relationships around culture, food, and people in the Southern Black Belt (the place of its birth), then it must undertake the valiant work of going back to these very uncomfortable truths to explore and expose how enslaved Africans and African Americans contributed to American culture and the food system. For the purpose of this discussion, African American refers to "an American of African and especially of Black African descent" (*Merriam-Webster*). This discussion further narrows focus to African Americans whose ancestors have resided in the United States for several generations (Whitehead & Williams-Forson, n.d.)

It is difficult to reconcile that the Black Belt South is an area riddled with diet-related diseases while simultaneously being home to some of the most coveted and relished foods known to humankind. Fried chicken, sweet potatoes, collard greens, rice with gravy, okra, and black-eyed peas; the list of inherently delicious southern delicacies does not stop here. A true Southern summer promises wild honeysuckle, Gulf shrimp, fresh muscadines, peaches, watermelons, and if you are lucky, a glass of sweet tea. Many have termed these dishes and the culture that surrounds them as "soul food." Yet in the face of federally mandated nutrition programs, such as USDA's MyPlate, the questions beckon: Where do "Big Mama's" greens fit on my plate? Do Granddaddy's ribs have a seat at the table of our current food system?

MyPlate is the current nutrition guide published by the USDA Center for Nutrition Policy and Promotion and depicts a place setting with a plate divided into five food groups accompanied by a drinking glass (USDA, 2020b). MyPlate replaces the USDA MyPyramid guide as the USDA's healthy eating communication initiative (Levine et al., 2012). Researchers have posited that although the government website may seem successful at being "objective," in reality "their documents may marginalize or misrepresent certain cultures in the Unites States" (Bancroft, 2012, p. 17). In her review of the MyPlate website Bancroft finds,

> The USDA omits commonly eaten foods from other cultures, which excludes those minorities and adds to the layers of systemic discrimination they already face. . . . The USDA assumes what kinds of foods and dishes readers are already consuming, but the assumptions often are inappropriate for Asians, African Americans, and Latinos. (Bancroft, 2012, p. 73)

In this process of omitting the foods of minority groups and centering the foods of dominant groups, outreach programs can entrench a narrative within the food system that prioritizes the cultural values of dominant groups at the expense of minority groups. Thus, Cooperative Extension must acknowledge its role in ensuring equity in terms of the cultural integrity of the food system. Research must expand in terms of the appropriate integration of the cultural aspects of food in relation to the formation of policy and public health measures such as MyPlate. Levine et al. (2012, p. S10) contend, "Two important areas warranting more research and observation are MyPlate's message use in vulnerable populations, and signs of its impact on social norms."

Beyond providing mere sustenance, foodways hold a tremendous cultural and physiological value to people. Ethnic groups maintain their cultural identity, values, and beliefs via foodways (Goody & Drago, 2009, p. 43). Thus, food is powerful as it both constructs and maintains social identity. Throughout time, people made decisions about what they ate based on cultural values; thus food is representative of culture transmitted through history. Food holds the history of a given ethnic group and the processes they transverse. To know a people is to know their food.

Moreover, food is not only about consumption and identity; food is about power (Alvarez, n.d.). This is especially true for the South, a place revered for its hospitality, a reputation born of the work and toil of enslaved Black chefs and farmers (Deetz, 2018). Far from harmless, food played a vital role in the safeguard of enslavement and white supremacy in America. Food was a powerful and painful method of control in the hands of enslavers while simultaneously playing a foundational role in the resilience of enslaved peoples who utilized it as a tool of resistance.

Understanding the history of food and its complex interactions in different contexts can reveal a better understanding of the current values and outcomes associated with the integration of cultural and ethnic foods into the system.

From the slave codes to the Selma to Montgomery march, food represents a form of resistance and continuation for Black people living in the South. There are tales of plants like okra, the beloved ingredient in the southern delicacy "gumbo," traveling from Africa to America in the bowels of slave ships, hidden in the braided scalps of stolen African women who brought these seeds in hopes of ensuring their own survival in whatever foreign and cruel place they would land (Blalock, 2019). We cannot forget that the same grits once eaten early morning in Southern fields filling sharecroppers' stomachs now find their way onto the menus of five-star restaurants. Not to mention the folklore that extolls an unnamed enslaved African woman believed by many to have taught her enslaver somewhere on the shores of South Carolina's coastal islands how to successfully cultivate rice, thus helping to bring the industry to the United States (Holloway, 2006). The integration of "soul food" or "Southern food" or "Black folks' food" into the system has a torrid history of extractions and neglect.

What Is Soul Food? A Personal Reflection on a Nuanced Conversation

To begin this section, author Lindsey Lunsford finds herself compelled to briefly offer personal reflections on her own intimate connections to food and history: My story reflects that of a Cooperative Extension professional, a resident of the Black Belt, and Black woman working to reconcile the inequities within the land-grant system and my role in ensuring equity in the food systems I serve. If the adage is true that "you are what you eat," then the best parts of me came from my grandmother. The warmth, the meticulousness, the careful preparation: this is how she made her food, and she was not alone. Somewhere in history, tucked away as tidily as my grandmother's kitchen recipes, sits the long history of African American women and men whose mastery of culinary arts and experiences led to the making of generations of freedom seekers and modern-day strivers.

The culinary history of African Americans is no different than that of American history as it entangled with the vestiges of enslavement, imprisonment, race, class, and gender struggles (Harris, 2011). For as sweet and as warm as my grandmother's rolls were, they could not escape their direct lineage from chattel slavery no more than my grandmother or I.

African Americans across the United States are attempting to rewrite and reclaim our ancestral heritage of traditionally healthy food systems, often explicitly advocating for food justice and equity. As Leah Penniman (2018, p. 8) writes, "Owning our own land, growing our own food, educating our own children, participating in our own health care systems—this is the source of real power and dignity." By reclaiming the narratives surrounding African American food systems and foodways, we are restoring our senses of self, our cultures, traditions, and legacies. Conducting this research as a Black woman, I aim to contribute to the African American food justice movement through the reclamation of our food system story. Most specifically, I seek to challenge and reclaim narratives around the social construct known as "soul food" or African American heritage cooking, as a cultural representative for African American food systems and foodways. For all marginalized and oppressed people, the recovering and reclamation of our food system story is integral to achieving self-determination, economic and cultural prosperity, and public health (Arthur & Porter, 2019).

So what then is soul food? According to culinary historian Michael Twitty (2017, p. 5), it is the "most remarked and most maligned of any regional or indigenous ethnic tradition in the United States." Famed African American foodways researcher Dr. Jessica Harris (2011, p. 208) deems soul food "a combination of nostalgia for and pride in the food of those who came before." Although popularly personified as greasy, high-fat, and calorically dense fried foods, soul food, like most phenomena, is more encompassing and diverse than as personified in mainstream media advertisement. Ranging from the rations of hog meat and hominy given to enslaved African Americans in the South to the vegan soul food prepared in Oakland by Chef Bryant Terry, soul food is as diverse as those that make and consume it (Harris, 2011). As Reese and Garth explain, "When we look closely at the literature, we find that there is a whole lesser-known history of teaching 'industry and self-reliance' in the kitchen, using fresh fruits and vegetables and maintaining a variety of nutrients in Black cuisine 'that shunned frying or dependence upon fatback seasoning'" (Reese & Garth, 2019, para. 6). Although soul food is widely diverse, some tenets persist as historian Frederick Douglass Opie (2008, p. 1) explains: "African American cuisine—what African Americans in the 1960s would later call 'soul food'—developed from a mixing of the cooking traditions of West Africans, Western Europeans, and Amerindians." Although influenced by other cultures, soul food is inherently the "intellectual invention and property of African Americans" (Opie, 2008, p. xi). Cooperative Extension has a role to play in promoting the cultural integration of soul food as it relates to the values of African American food systems.

Case Study: Tuskegee University Cooperative Extension Program's Annual "FEASTIVALS"

Massive marketing and food delivery services are transforming the diets of all Americans (Whitehead & Williams-Forson, n.d.). Moreover, as the diets of African Americans have continued to diversify over time and include broader and more diverse ingredients and grow further from historic or traditional core African American diets, feast and festivals have remained a way to preserve and uphold traditional African American foodways such as soul food. Drs. Williams-Forson and Whitehead state:

> The feasts that bring together African Americans, particularly those with cultural roots in the South, provide continuity in African American traditional core foods. As long as festive occasions and the rituals that accompany them continue, those foods will continue as well. (Whitehead & Williams-Forson, n.d., para. 53)

An example of a Cooperative Extension program using festive occasions to help preserve traditional African American food pathways is the Tuskegee University Cooperative Extension (TUCE) "FEASTIVAL." The concept initiated in 2014 as a way to welcome community members to a newly opened community garden project (Carlisle, n.d.). Since its inception, the FEASTIVAL has grown to feed over two thousand residents living in the Black Belt region (Macon County, AL). The event utilizes local chefs and elder women and men of the community whose collective memories offer insights into the preparation and handling of traditional African American core foods. In the traditional nature of soul food, the program relies on crops grown in season at TUCE-sponsored community gardens for the preparation of community meals. Such meals include fresh collard greens, mashed sweet potatoes, stewed okra, salad, sautéed squash, and other fruits and vegetables. The incorporation of dark leafy greens, sweet potatoes, okra, and watermelon evoke the use of centuries-old foodways and crops with their origins in the African continent (Carney & Rosomoff, 2009).

According to Huang et al. (2007, p. 1396), "Collard greens, mustard greens, kale, okra, sweet potato greens, green onion, butter beans, butter peas, purple hull peas, rutabagas, eggplant, and purslane are commonly consumed and considered as traditional foods by African Americans." Research suggests that "these indigenous vegetables among African Americans" are good sources of nutrients that can be useful for the prevention of cardiovascular and other chronic diseases (Huang et al., 2007, p. 1401). By utilizing crops that have a cultural and historical relevance to the region and its inhabitants, the FEASTIVAL encourages and supports traditional African American foodways.

The event's title, "FEASTIVAL," includes the extra "A" to remind the attendees of the role of arts and agriculture in the festival as well as to highlight the importance of "feast" in the festivities. Attendees enjoy music, social gathering, and culturally and ethnically appropriate foods prepared by Extension professionals and local chefs. By gathering African Americans in the Black Belt around joyful events that honor the significance and role of fresh local and culturally and ethnically appropriate foods, TUCE is engaging in the noble work of preserving traditional African American foodways for generations to come.

Conclusion

This chapter examines the U.S. Department of Agriculture and the land-grant system and finds them wanting in their commitment to serve a wide swath of land called the Black Belt South, and a large section of the population, rural African Americans. At the same time, there exist land-grant institutions, the 1890s, that, despite being underfunded, provided these same people with outreach and Extension programs. The importance of this outreach was so significant that formal Cooperative Extension activities, that is, a partnership between the USDA and land-grant institutions, began at Tuskegee in 1906 with the appointment of T. M. Campbell, a full eight years before the Smith-Lever Act.

The full thrust of 1890 Cooperative Extension, and research, has been impeded by legislative action so that independent funding for the historically Black land-grants did not occur until well into the latter half of the twentieth century while, currently, federally mandated state matching funds still elude many 1890 universities. At the same time, however, the commitment of formal Extension programs, as well as additional programs such as 2501, is still strong. Indeed, it has become a necessity for 1890 universities to tap into federal funding to help support their programs, and to promote more action and support for these programs.

Finally, in terms of cultural integrity we question what becomes of the role of Cooperative Extension in acknowledging the history of rural African American foodways and in preserving and promoting their value in the current food system. The narrative must resurface that centers Cooperative Extension's birthplace in Tuskegee, Alabama, the Black Belt South, in the hands and hearts of African American people. Thus, it becomes the call, more than a century later, to deliver financial, systematic, and cultural equity to the land, people, and customs that led to the creation of the land-grant system under the care of the USDA.

REFERENCES

Alvarez, L. (n.d.). Colonization, Food, and the Practice of Eating. Food Empowerment Project. https://foodispower.org/our-food-choices/colonization-food-and-the-practice-of-eating/.

Arthur, M., & Porter, C. (2019). Restorying Northern Arapaho Food Sovereignty. *Journal of Agriculture, Food Systems, and Community Development*, 9(B). https://doi.org/10.5304/jafscd.2019.09B.012.

Bancroft, J. (2012). The USDA Food Plate Website: Culturally Conscious or Colorblind? Master's thesis, Texas State University-San Marcos.

Blalock, B. (2019). African-Americans Have Shaped Alabama's and America's Cuisine. Alabama News Center, February 6. https://alabamanewscenter.com/2019/02/06/african-americans-shaped-alabamas-americas-cuisine/.

Carlisle, J. (n.d.). 3rd Annual TULIP Feastival Observed. *The Tuskegee News*. Retrieved December 4, 2019, http://www.thetuskegeenews.com/news/rd-annual-tulip-feastival-observed/article_4e17b216-2a91-11e7-b482-031c00b2fc1c.html.

Carney, J. A., & Rosomoff, R. N. (2009). *In the Shadow of Slavery*. Berkeley: University of California Press.

Deetz, K. (2018). How Enslaved Chefs Helped Shape American Cuisine. *Smithsonian Magazine*, July 20. https://www.smithsonianmag.com/history/how-enslaved-chefs-helped-shape-american-cuisine-180969697/.

Executive Order 12320. (1980). Historically Black Colleges and Universities. National Archives. https://www.archives.gov/federal-register/codification/executive-order/12320.html.

Executive Order 12876. (1993). Historically Black Colleges and Universities. https://www.govinfo.gov/content/pkg/WCPD-1993-11-08/pdf/WCPD-1993-11-08-Pg2234.pdf.

Goody, C. M., & Drago, L. (2009). Using Cultural Competence Constructs to Understand Food Practices and Provide Diabetes Care and Education. *Diabetes Spectrum*, 22(1), 43–47. https://doi.org/10.2337/diaspect.22.1.43.

Davis, J. (2000). Social Stress and Mental Health in the Southern Black Belt. *Sociological Spectrum*, 20(4), 465–494. https://doi.org/10.1080/02732170050122648.

Hargrove, T. M. (2002). A Case Study Analysis of the Implementation Process of the Small Farmers Outreach Training and Technical Assistance (2501) Program, 1994–2001: Implications for Agricultural Extension Education Targeting African American Farmers. PhD dissertation, Iowa State University. http://lib.dr.iastate.edu/rtd/376.

Hargrove, T. M., & Jones, B. L. (2004). A Qualitative Case Study Analysis of the Small Farmers Outreach Training and Technical Assistance (2501) Program from 1994–2001: Implications for African American

Farmers. *Journal of Agricultural Education*, 45(2), 72–82. http://www.jae-online.org/attachments/article/305/45-02-072.pdf.

Harris, J. B. (2011). *High on the Hog: A Culinary Journey from Africa to America.* New York: Bloomsbury Publishing USA.

Holloway, J. E. (2006). African Crops and Slave Cuisine. Slave Rebellion Web Site.http://slaverebellion.info/index.php?page=crops-slave-cuisines.

Huang, Z., Wang, B., Eaves, D. H., Shikany, J. M., & Pace, R. D. (2007). Phenolic Compound Profile of Selected Vegetables Frequently Consumed by African Americans in the Southeast United States. *Food Chemistry*, 103(4), 1395–1402. https://doi.org/10.1016/j.foodchem.2006.10.077.

Land Grants. (2019a). First Morrill Act. http://www.higher-ed.org/resources/morrill1.htm.

———. (2019b). Second Morrill Act. http://www.higher-ed.org/resources/morrill2.htm.

———. (2019c). Hatch Act. http://www.higher-ed.org/resources/hatch.htm.

———. (2019d). Smith-Lever Act. http://www.higher-ed.org/resources/smith.htm.

Lee, J. M., Jr., & Keys, S. W. (2013). *Land-Grant but Unequal: State One-to-One Match Funding for 1890 Land-Grant Universities.* Washington, DC: Association of Public and Land-grant Universities. https://www.aplu.org/library/land-grant-but-unequal-state-one-to-one-match-funding-for-1890-land-grant-universities/file.

Levine, E., Abbatangelo-Gray, J., Mobley, A. R., McLaughlin, G. R., & Herzog, J. (2012). Evaluating MyPlate: An Expanded Framework Using Traditional and Nontraditional Metrics for Assessing Health Communication Campaigns. *Journal of Nutrition Education and Behavior*, 44(4), S2–S12. https://doi.org/10.1016/j.jneb.2012.05.011.

Lincoln, A. (1864). Fourth Annual Message. December 6. https://www.presidency.ucsb.edu/documents/fourth-annual-message-8.

Mayberry, B. D. (1989). *The Role of Tuskegee University in the Origin, Growth and Development of the Negro Cooperative Extension System 1881–1990.* Montgomery, AL: Brown Printing Company.

Opie, F. D. (2010). *Hog and Hominy: Soul Food from Africa to America.* New York: Columbia University Press.

Pennick, E. (n.d.). *African-American Land Retention, Acquisition and Sustainable Development: Eradicating Poverty and Building Intergenerational Wealth in the Black Belt Region. Southern Regional Asset Building Coalition.* Tuskegee, AL: Tuskegee University/Ford Foundation.

Penniman, L. (2018). *Farming While Black: Soul Fire Farm's Practical Guide to Liberation on the Land.* White River Junction, VT: Chelsea Green Publishing.

Reese, A., & Garth, H. (2019). Beyond the Parentheticals: The Practice of Being in Conversation (Response to "From the Editor: The Politics of Citation in the Field of Food Studies"). *Food Stuffs*, 6(1), June 16. https://gradfoodstudies.org/2019/06/16/beyond-the-parentheticals/.

Rickenbach, M., Mohamed, A., Norland, E., & Blanche, C. (2013). Final Report: Review of McIntire-Stennis Cooperative Forestry Research Program. United States Department of Agriculture. June 3. https://www.fs.fed.us/research/docs/forestry-research-council/articles/mcintire-stennis-cooperative-forestry-research-program-review.pdf.

Schor, J. (1986). The Black Presence in the U. S. Cooperative Extension Service since 1945: An American Quest for Service and Equity. *Agricultural History*, 60(2), 137–153.

Seals, R. G. (1986). The Disparity in Land Grant Funding from State Sources between Traditionally White Institutions and Traditionally Black Institutions in Louisiana and Mississippi: The Role of Agricultural Development Legislation. *Journal of Rural Studies*, 2(3), 221–232.

———. (1998). *Disparity: An Analysis of the Historical, Political, and Funding Factors at the State Level Affecting Black Academic Agriculture*. New York: Vantage Press.

Tullos, A. (2004). The Black Belt. Southern Spaces. https://southernspaces.org/2004/black-belt/.

Tuskegee University Archives. Frederick D. Patterson Papers.

Twitty, M. W. (2017). *The Cooking Gene: A Journey Through African American Culinary History in the Old South*. New York: HarperCollins.

USDA (U.S. Department of Agriculture). (2001). Outreach and Assistance for Socially Disadvantaged Farmers and Ranchers Program History [News Release].

———. (2018). National Institute of Food and Agriculture. 7 CRF Part 3419. RIN 0524-AA68. Matching Funds Requirements for Agricultural Research and Extension Capacity Funds at 1890 Land-Grant Institutions, Including Central State University, Tuskegee University, and West Virginia State University, and at 1862 Institutions in Insular Areas. *Federal Register*, 83(92), 21846–21849. https://www.govinfo.gov/content/pkg/FR-2018-05-11/pdf/2018-10015.pdf.

———. (2019a). Outreach and Assistance for Socially Disadvantaged Farmers and Ranchers and Veteran Farmers and Ranchers Program (The 2501 Program). FY 2019 Synopsis of Funded Projects. https://www.usda.gov/sites/default/files/documents/fy2019-2501-grant-projects.pdf.

———. (2019b). USDA Announces $16 Million Funding Opportunity to Support Socially Disadvantaged and Veteran Farmers and Ranchers. Press Release No. 0110.19, July 16. https://www.usda.gov/media/press-releases/2019/07/16/usda-announces-16-million-funding-opportunity-support-socially.

———. (2020a). FY 2020 Allocation and Matching. https://nifa.usda.gov/resource/fy-2020-allocation-and-matching.

———. (2020b). Welcome to MyPlate. ChooseMyPlate. https://www.choosemyplate.gov/.

USDA (United States Department of Agriculture)/NIFA (National Institute of Food and Agriculture). (2019). Agricultural Extension at 1890 Land-Grant Institutions. 2020 Request for Applications. https://nifa.usda.gov/sites/default/files/resources/fy-2020-agricultural-extension-at-1890-land-grant-institutions-rfa-20190815.pdf.

Washington, B. T. (1901). *Up from Slavery*. New York: Doubleday & Company.

Whitehead, T. L. (1992). In Search of Soul Food and Meaning: Culture, Food, and Health. In H. A. Baer & Y. Jones (Eds.), *African Americans in the South: Issues of Race, Class, and Gender* (pp. 94–110). Athens: University of Georgia Press.

Whitehead, T. L., & Williams-Forson, P. (n.d.). African American Foodways. Encyclopedia.com. https://www.encyclopedia.com/food/encyclopedias-almanacs-transcripts-and-maps/african-american-foodways.

Wimberley, R. C., & Morris, L. V. (1997). *The Southern Black Belt: A National Perspective*. Lexington, KY: TVA Rural Studies.

Wimberley, R. C., Morris, L. V., & Bachtel, D. C. (1991). Agriculture and Life Conditions in the Land-Grant Black Belt: Past, Present, and Policy Questions. In N. Baharanyi, R. Zabawa, A. Maretzki, & W. Hill (Eds.), *Public and Private Partnership for Rural Development* (pp. 33–48). Tuskegee, AL: Tuskegee University.

Wimberley, R., Morris, L., & Harris, R. (2014). A Federal Commission for the Black Belt South. *Professional Agricultural Workers Journal*, 2(1). https://tuspubs.tuskegee.edu/pawj/vol2/iss1/6.

Reaching Marginalized Audiences through Positive Youth Development Programming: Challenges When Influencing Constituents Do Not Agree that 4-H Should Be Open to All

Jeff Howard and Amanda Wahle

The 4-H youth development program is America's largest youth development program, empowering nearly six million youth across the United States with skills to lead for a lifetime (National 4-H Council, 2019). The 4-H program started with the hope of knowledge transfer, through youth–adult partnerships, in an effort to get new agricultural technology into the hands of reluctant farmers. A. B. Graham is credited with starting one of the first 4-H clubs in 1902 in Ohio. The club allowed youth to concentrate on growing tomatoes and corn. Youth were taught different techniques, by the adults, to grow and tend to crops, which ended up being very successful for them. The hope in teaching youth was that they would then take these lessons home to their families to help with their own crops, thus transferring the lessons learned (Borden, Perkins & Hawkey, 2014).

With these basic grassroots approaches as a beginning, 4-H has evolved forward to provide high-quality positive youth development programming. During the decade of the 2000s, National 4-H conducted a ten-year study of the effectiveness of its program. The first-of-its-kind study demonstrated that 4-H was uniquely postured with its hands-on approach to give kids the opportunity to learn by doing and building life skills. The study included over seven thousand adolescents from diverse backgrounds in forty-two states. The study demonstrated that 4-H has evolved to create a construct where youth who engage are four times more likely to give back to their communities, two times more likely to make healthier choices, and two times more likely to participate in science, technology, engineering, and mathematics, which may create more opportunity for them in contemporary times (Lerner, Lerner & Almerigi, 2005).

"The 4-H Motto, 'To Make the Best Better,' encapsulates the focus of the program, namely empowering youth with the tools and knowledge to reach their full potential, while working and learning in partnership with caring adults" (Borden, Perkins & Hawkey, 2014). Nonformal education programs, such as 4-H, can be excellent companion programs with schools and after-school programs.[1] The 4-H model of do, reflect, apply is experientially focused so children retain information much better if they have experienced it firsthand and applied the new information they've gained from the experience into their daily lives. Curriculum that is peer reviewed and approved for the 4-H name and emblem is experientially focused and requires points of rigor that assure experiential development is a core curricular component. Comparative studies have found that classrooms that have incorporated 4-H into learning plans demonstrated higher levels of standardized test scores and lower levels of absences and tardiness (Klemmer & Zajicek, 2002).[2]

In 1914, the Cooperative Extension Service was created with the passage of the Smith-Lever Act allowing for a more nationalized approach to the 4-H program. Cooperative Extension evolved over the decades to be a partnership with what is now known as the National Institute of Food and Agriculture (NIFA) within the U.S. Department of Agriculture (USDA).[3] This partnership grew to include more than one hundred land-grant universities and over three thousand local offices nationwide (National 4-H Council, 2019). This partnership between universities and government agencies sets the 4-H program apart from other youth-serving organizations in that it is funded through a variety of resources from local, state, and federal governments along with expertise to provide programming to youth in our communities. It is often the only connection that youth may have with a university-based educational program within their community. This exposure to a university-focused program may be a conduit to interest in further pursuit of an education, thus improving a youth's potential for success, earning, and contribution. "Research-based best practices to improve college attendance include college awareness experiences, mentoring programs, and parent and peer involvement. As an outreach of land-grant universities, 4-H programs are well positioned to help adolescents with college preparation and access" (Copeland et al., 2009). But the benefit to all youth has been impacted by the historical roots of 4-H and its continued struggles with diverse populations. Membership in 4-H peaked in 1974 at 7.5 million (Van Horn et.al). Another significant way in which 4-H has changed is in integration of clubs (Van Horn, Flanagan & Thomson, 1999). Van Horn, Flanagan, and Thomson note that before the Civil Rights Act of 1964, some 4-H clubs were racially segregated and some clubs in schools were segregated until 1975 because schools remained segregated. Today, the 4-H program is committed to serving all youth within communities. Nonetheless, the stereotype of 4-H as an organization of white, rural youth has been difficult to overcome. Since most communities tend to be homogeneous and the club model is usually based within the community, developing a diverse club membership can sometimes be difficult.[4]

How 4-H Is Postured for Marginalized Audiences

Although 4-H has existed for more than one hundred years and served millions of children during that time frame, the reality is that it was predominantly focused on white children in rural settings and, as noted earlier, the perception may be that it is still is. There were some exceptions, however, such as the development of the 4-H program in North Carolina for African Americans in 1914 following the hiring of the first African American agricultural agent. Within a few years the program grew to include county camp opportunities and short courses. By 1936 there were ten thousand African American youth being served. With increased programming, and more opportunity, the growth of the African American program continued through the 1940s to serve thirty-six thousand youth by 1946 (Manor & Pronovost, 2007).

While the concentrated inclusion of African American youth is something to celebrate, other marginalized youth may not have received the same level of dedicated focus. Marginalized youth are often described in the youth development field of study as those children who are on the lower or peripheral edge of society who may be denied involvement in mainstream societal experiences. This exclusion could come as a result of economic factors, political influence or lack thereof, as well as cultural and social activities that they may have experienced challenges engaging in or as something where they are not welcomed. Further, youth in urban settings, whether marginalized or not, may not have perceived the program as something that had anything to offer them. Recognizing that these tensions exist, 4-H took action to broaden the program from only a community club experience in the 1970s to begin offering 4-H in partnerships with schools, aftercare programs, and community program partners.

As such, marginalized youth began to be intentionally included, although it is acknowledged that their connection was often principled on their involvement in the partnership programs. Marginalized youth began experiencing 4-H and had more consistent exposure to experiential learning, a foundational element of 4-H that can benefit all youth, marginalized children included. These elements were designed with the idea that regardless of what program is being offered or the venue that it occurs in, if these elements are a part of the planning and execution then the youth involved will more likely receive what they need for a quality experience and potentially positively influence their life skill development. The 4-H programming should be approached from the place of giving youth the foundation for growth, which then allows for learning to occur.

The 4-H program has taken intentional steps to create outreach opportunities beyond the club program, which enhances the probability that marginalized youth will experience 4-H. Through these opportunities they may experience elements of the program that are essential to their development, which are referred to as the Essential Elements of 4-H.

Belonging is the first of the four essential elements and creates the foundation for welcoming children into the program as well as keeping them involved year after year. The positive connection with a caring adult that is not from their immediate family is a key factor in the 4-H program. Children need someone in their life that they can trust, feel safe with, and go to beyond their family members. Through the 4-H program volunteer model those connections are developed through club leaders, activity leaders, Extension educators, coaches, trip chaperones, camp counselors, and so many more. Once a connection is made with a caring adult the environment will then become a place where the child may feel safe. Programmatic environments for 4-H, whether club meeting, after-school program, camp facility, or larger statewide event, need to be places where the youth involved feel safe and included. An inclusive environment is a strong component of developing a sense of belonging, the idea that everyone is welcome and will be provided an equal opportunity to grow and learn in the program.

The second of the essential elements is mastery. Mastery is the idea that youth are given opportunities to engage with a program or project area in a way that allows for mastering of skills. Mastery doesn't always mean best in show or champion ribbons; for some youth it means hitting an archery target after developing techniques for holding a bow, or completing a project that can be entered in the fair. The goal with mastery is continued growth and learning and also being able to encourage new challenges when something has become too easy. Mastery is also a great way to encourage goal setting and the development of small goals that lead to successful completion of a task.

Independence, the third essential element, goes along with mastery with the thought that every child needs to feel like they are an active participant in what is happening with the club or group. They need to see themselves as a valuable part of the process and someone that is contributing. With independence comes the hope that youth will have the opportunity for self-determination. Through 4-H, opportunities for self-determination are created through project area choice, offering different focus areas for clubs and state and national contests in different program areas, and by continuing to develop and grow program areas based on need and interest. The 4-H program is designed so that youth focus on their area of interest and grow from there.

The final essential element is generosity. The idea of generosity is youth looking beyond themselves and seeing the value in providing service to others. This service can come in many forms, but it really is about giving back in a way that makes the youth contributing and successful members of their community. Providing leadership, setting good examples for younger members, advocating for a cause, or cleaning up trash in community areas are all ways that 4-H members have given back.

In 2018 an updated model was developed by Oregon State University and is known as the Thrive Model (Arnold & Gagnon, 2020). It was developed to enhance the demonstrated needs of youth and methods to get them engaged and keep them engaged with the 4-H program.

The idea of this model is if quality 4-H programming is provided by Extension and its volunteers then youth will thrive. Youth who thrive achieve key development outcomes such as academic achievement and motivation, social competence, increased personal standards, contribution and connection to others, and personal responsibility. The first area, which is similar to the model of essential elements, is centered on youth sparks. The 4-H program needs to be a place that allows youth to explore their interests and passions. This interest combined with a quality positive youth development program and the opportunity for developmental relationships with caring adults starts the process for a youth to thrive in the 4-H program.

Working through the model will provide youth with the opportunity to try new challenges, develop a growth mindset, have a belief in a hopeful future and purpose in life, connect to a

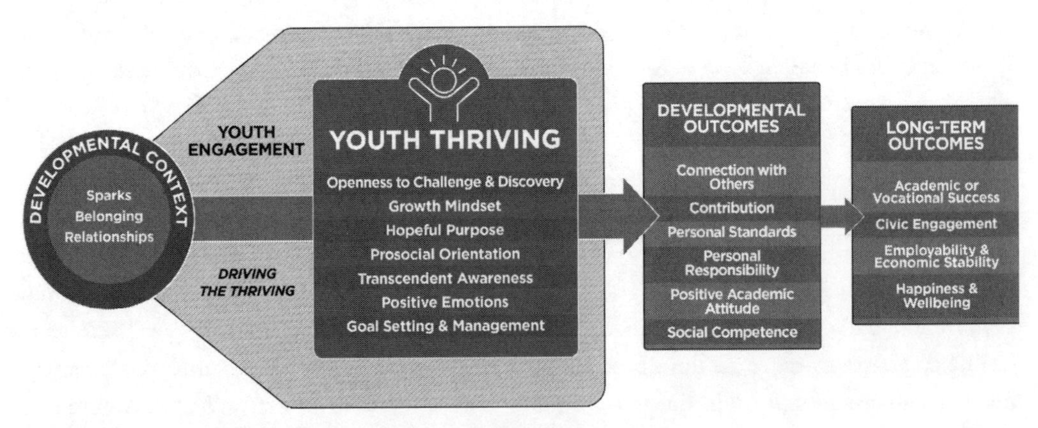

FIGURE 1. The 4-H Thriving Model (Arnold & Gagnon, 2019).

greater sense of self, manage emotions, possess and express values, and set and manage goals (Arnold & Gagnon, 2020). Being able to develop through the program and gain these skills will lead to the developmental outcomes that are hoped for in the program.

The power behind both the essential elements and the Thrive Model (figure 1) is that the development of one area leads to the rest unfolding. These models are meant to be guides in all work that is done in the 4-H program so that they naturally fall into place as people join and grow as members. These models develop a foundation for a program that can be open and accepting to all youth, and as that happens marginalized youth that have been included will benefit as well.

Challenges Experienced Including Marginalized LGBTQ Youth in 4-H

Although the 4-H program is designed, in theory, to benefit marginalized youth, there have been struggles. Historically, there may have been times when recommended practices have not aligned with societal acceptance. To reiterate, the actual origins of the 4-H program are tied to an educator who chose to focus on white boys in rural communities to teach agriculture production practices. Although the model advocates for acceptance of all, there were times in our past when the evolution of 4-H may not have been in alignment with perspectives of how the program should be structured. Considering the evolution of the civil rights movement, it is reasonable to imagine the opposition that was a societal norm within schools and communities and a similar perspective that 4-H should continue as a program for rural boys who were white. A truly telling observation of the separated Extension programs for white and Black youth is found in the book *Bittersweet Perspectives on Maryland's Extension Service* (Joelle & Lofton, 1990).

The separated opportunities and inequitable resources brought public controversy. However, Extension workers pressed the program forward to embrace diversity and richen the cultural depth that our youth may experience through their club. Foundational antidiscrimination policy is rooted in the 4-H experience, which one would theorize would lead to a broader level of acceptance of all. However, as we aspire to become more culturally competent, other societally identified forms of diversity have come to the forefront that have met resistance from some within 4-H.

The 4-H program has developed intentional approaches to focus on access, equity, and belonging with a goal of increasing supportive and inclusive programs.[5] While there may have been an enhanced awareness and advocacy to create more belonging environments for marginalized youth, an element of diversity that has had less deliberate attention in the past is acceptance and awareness of youth who are part of the lesbian, gay, bisexual, transgender, and queer (LGBTQ) community. Youth are "coming out," a term regarding a youth sharing their sexual orientation with others, at much earlier ages as compared to previous decades (American Academy of Pediatrics, 2018).[6] As such, 4-H workers must become more knowledgeable about the LGBTQ child and how to best support them, especially when they might not receive support in more ideologically conservative settings. Furthermore, youth may come out to members of their 4-H club, but that does not mean our clubs, volunteers, and educators are ready to receive them. In a nationwide survey conducted with Extension administrators and youth workers in 2017 based on observations, responses from thirty-one states found that only 17 percent observed their communities to be supportive of the LGBTQ community while 31 percent indicated that they observe tolerance. Lastly, 10 percent of the respondents indicate observations of hostility.

Extension workers identified the need to receive more training in this topical area as well as the continued expressed need for more support materials to help educate them so they can better support LGBTQ youth and families (Howard, 2019). Although all youth are at some risk, youth within vulnerable populations may have an increased level of risk factors. Youth who are at risk need to experience belonging to a community that supports them (Hensley et al., 2007). The 4-H club experience has provided a venue for LGBTQ children to have positive influences that are essential to their development. While all youth need inclusive environments, the issues that LGBTQ youth face are unique (Howard, 2019). Whether a child is "out" to their 4-H club or not, the sense of belonging, through 4-H, and having significant noncustodial adults in their lives can be positive. The 4-H club provides a venue where a significant adult in a child's life may be regarded as a mentor or role model that LGBTQ youth may not find otherwise, thus decreasing their social capital. Acceptance influences a child's self-esteem and social support, thus 4-H can provide an environment where children feel accepted and they belong. Consider also that in 4-H the program teaches youth to debate in a civil manner and speak in public, as well as teaching informed decision making and community engagement. These experiences may enhance a child's confidence and decrease risk factors, which all children need, but LGBTQ youth even more so as they may face more adverse challenges in their lives than many in their peer set.

As youth development professionals, there are easy approaches that can be taken to demonstrate a more inclusive environment for LGBTQ youth and families. Dr. Katherine Soule outlines a number of actions that can be taken to demonstrate inclusion (Soule, 2017). A basic focus is to educate one's self regarding language. Language matters and can be a subtle way to demonstrate support for LGBTQ youth. The use of inclusive and accepting language will go far for LGBTQ families to trust the program has their child's best interest at heart. As promotion materials are developed for 4-H programming, 4-H workers should highlight the program is accepting of *all* youth. A youth worker can display supportive stickers or signage that demonstrates their role as an ally. Workers in 4-H can evolve their registration materials to ask for "parents" rather than "mother and father," as one considers it will continue to become more common that children may have same sex parents or other familial structure beyond a traditional nuclear family. Similarly, ask for "gender" on forms rather than "boy or girl." These subtle adjustments could demonstrate to LGBTQ individuals and families that the 4-H program values them and offers a place for them to belong.

Case Study: Growing 4-H when Inclusion of All Is in Opposition to Administration and Constituent Groups

During 2014, with a goal of better aligning the 4-H worker with understanding and supporting youth within more vulnerable populations, the National 4-H Headquarters led an effort across the 4-H movement to establish working teams populated with Extension professionals from across the nation that focused on a variety of vulnerable populations. Teams were formed to support youth including those in foster care, immigrant youth, youth with disabilities, and migrant youth, to name a few. One particular work team focused on the development of curriculum and workshops concentrated on developing understanding of LGBTQ youth.

The LGBTQ Vulnerable Population Workgroup was authorized by the National 4-H Headquarters and, under agreed-upon leadership, began their work including a national assessment of levels of understanding at that time and identifying needs regarding curriculum and workshop support. The work team identified a series of workshop needs and began to submit proposals, focused on developing inclusion and acceptance, at regional- and national-level professional meetings. The team also partnered with the National 4-H Headquarters to host a nationwide webinar as part of an educational series titled "4-H eAcademy: Increasing Inclusivity for Youth Who Identify as Members of LGBTQ+ Communities." Between 2016 and 2019, members of the work team presented thirteen workshops at regional, national, and international levels.

In concert with the workshop series and various national presentations, the team also worked on a support document that would focus on guiding principles for LGBTQ acceptance and inclusion that states could choose to use as they developed their own state's policy documents at each land-grant university. Simultaneously, a group in the Western Extension region of the United States developed a document that was based on policy from California Extension. The national work team focused a vetting session in the summer of 2017 on the existing document and ultimately decided to endorse the document as an example that land grants could draw on. Furthermore, the work team felt such a level of confidence in the prepared support piece that it was sent forward to the National 4-H Headquarters as a curricular piece to make available to Extension 4-H workers as well. To the work team's knowledge, protocols in place at the National 4-H Headquarters were followed regarding review and acceptance of the document, which eventually led to the document being authorized for use.

On March 13, 2018, a courtesy note was sent out nationwide to alert Extension programs that the document was available for utilization as needed by states via the National 4-H Headquarters' website where the documents produced by the vulnerable populations work team were housed. Simultaneous to the nationwide notification, social media traffic began to escalate from some constituent groups focused on denouncing not only the guiding principles document, but also the inclusion of LGBTQ youth in 4-H and some of the suggested elements of programming that would help the youth to feel accepted and that they belonged. The constituent groups in opposition concentrated on language that 4-H was "pushing the gay agenda" and that the constituent group's religious freedoms were being affected. As media postings escalated, 4-H members, parents, supporters, and volunteers who were supportive of the inclusion guidance began to share supportive responses, and debate and arguments continued to populate social media. Based on chapter author Howard's experience as a State 4-H program leader, the guiding principles document was highly sought after by a multitude of states and youth workers. However, within days of the posting, the document was removed and an electronic mailing was sent to all State 4-H program leaders, including the author, from an official at the National Institute of Food and Agriculture including the following language: "On March 13, 2018, the National Institute of Food and Agriculture (NIFA) circulated an email containing a link to a document developed by a regional 4-H leadership group. The document discussed material unrelated to the mission of 4-H."[7] The email was vague, and although the specific connection was not highlighted in the brief note, it was referencing the courtesy note announcing the availability of the document "Practices for Inclusion of Individuals of All Genders and Sexual Orientations" (Program Leaders Working Group, 2020). The national LGBTQ Vulnerable Population Workgroup was not consulted or directly notified of the decision, and following repeated contact for an explanation, no further explanation was provided.

At the request of the LGBTQ Vulnerable Population Workgroup, the national Extension Committee on Organization and Policy (ECOP) sent an inquiry to the deputy director of NIFA seeking an explanation regarding the removal of the guiding principles document and related support documents from the website. The response received indicated that the document had not followed the proper vetting process to obtain a USDA logo, which led to its removal.

The LGBTQ Vulnerable Population Workgroup continued to push for affirmation that their work could continue, but having never received a commitment to continue forward, the team, as well as other vulnerable populations work teams, advocated creating a new structure, led by faculty at selected land-grant universities. The new dynamic was supported by ECOP, and the Access, Equity, and Belonging chartered process was begun.

During the summer of 2018, several other significant issues caused a national level outcry from opponents and supporters of LGBTQ inclusion in 4-H. A State 4-H program leader was dismissed from his job, without notice, following the program leader's advocacy for the adoption of the LGBTQ guiding principles document as policy (Crowder, 2018). During this time frame, the Ohio State 4-H program hosted a national LGBTQ Youth Summit that prompted protest to 4-H as well as celebration of inclusive workshops and training. Based on postings on social media, support for LGBTQ inclusion was observed from 4-H alumni, parents, members, and Extension faculty from around the nation. However, constituents drew on social media to share their opposition to inclusion of LGBTQ youth in 4-H as well. The concern of a negative image of 4-H resulting from perceptions that 4-H was not actually open to *all* youth led to a national statement being issued by the National 4-H Council, the national 4-H private partner, affirming their commitment to all children and the inclusion of all youth in the 4-H program. Furthermore, ECOP issued a nationwide statement of support for LGBTQ youth inclusion cosigned by the CEO of the National 4-H Council:

> As Cooperative Extension Directors/Administrators, we encourage you to engage in this critical conversation. Our opportunity to be consistent in our commitment to diversity and inclusion in all Extension programming is now. In addition, we encourage you to support your 4-H leaders, workers and volunteers who may be challenged in their efforts to operate 4-H programs in the most inclusive manner possible, consistent with their respective Land-grant University policies. We also believe it is critically important to acknowledge and back up those who are willing to be champions in this cause.
>
> Cooperative Extension has an opportunity to support civic dialogue around issues of diversity and inclusion, and adhere to our mission of bringing evidenced-based information into the public arena to address these important issues. ECOP has authorized the ECOP 4-H Leadership Committee to continue these important conversations, respond to questions and concerns raised by our partners in 4-H, and ensure that as a national Cooperative Extension System we continue to embrace diversity and inclusion, ensure equity, and achieve our goal of a 4-H program that is reflective of the youth of our nation. (Extension Committee on Organization and Policy, 2019)

Although the support statement was issued, a news article published by the *Des Moines Register* alleged that the 4-H program leader's dismissal was because of his support of LGBTQ youth and the inclusion document (Crowder, 2018). University officials categorically denied this. The additional media coverage led to a letter being sent by a U.S. congressman, along with seventy additional congressional signatures, to the secretary of agriculture. The letter requested explanation regarding the guiding principles document being removed from the National 4-H Headquarters website as well as the concern of the perception that 4-H was discriminating against LGBTQ youth. A response came from the secretary indicating that the document was removed because it was not the USDA's role to issue policy for the land-grant universities to follow and that the document itself was not properly vetted via USDA protocol. The letter affirmed the USDA's commitment to civil rights, and the secretary underscored the statement of record regarding civil rights affirmation.

The case study presented leads to some significant questions that educators in Extension may face. The most grassroots question being "What does Extension do when programmatic support and needs of youth do not align with higher level administration and constituent interest and advocacy?" Further points to consider include organizational and program development approaches for Extension workers when perceived controversial topics or processes are in opposition to influential constituent groups. Another consideration is what approaches can be taken to push forward nonbiased basic human rights for all, when public outcry opposes progress.

The Trump administration, during 2018, demonstrated a number of actions that led to reduction in inclusion of all, including decisions regarding who could and could not serve in the military as part of the LGBTQ community. Research areas had also been affected, particularly regarding research surrounding climate change as another example of the influence of a sitting presidential administration. In a 2017 National Public Radio address attention was drawn to the intentional removal of climate change language to avoid funding cuts (Hersher, 2017). The fundamental reality of funding streams that may be tied to Extension's work can be affected if Extension continues work in an area that a sitting administration is in opposition to.

Extension workers will likely continue to face challenging situations concerning controversial topics and research opinions. Extension has navigated these situations for decades when one considers production practices that may go against historic norms or research findings that are in opposition to cultural norms of families. Cooperative Extension must develop more training, support pieces, and fact sheets to help Extension workers in their daily roles. Extension workers will need to be deliberate in hosting training workshops for volunteers surrounding this topic so 4-H club leaders have a base level understanding of the LGBTQ child. Resistance is to be expected, and a youth worker should be prepared for it. Although a civil/human rights issue, the reality is that LGBTQ acceptance has complex resistance, especially within something such as 4-H and Extension. Much of the resistance is based on religious and cultural beliefs. In 2020, seventy-two countries around the world still criminalize people in the LGBTQ community (Human Dignity Trust, 2020).

The prevalence of social media has created the reality that misinformation and advocacy for opposing negative opinions can be shared on a nationwide basis instantaneously. Extension workers will likely face multiple occurrences where the management of incorrect and inflammatory information becomes consuming to one's daily work. It is critical that dialogue and open conversations can be held regarding challenging topics so Extension educators are best postured to support all constituents. In the book *Jumping into Civic Life: Stories of Public Work from Extension Professionals*, a number of stories are shared that address this topic (Peters, Alter, & Shaffer, 2018). As Extension continues to advocate for cultural appreciation and competence, how does one navigate as a civic actor, when the advocacy may result in loss of support and funding from historically influential people?

When we consider programming as a whole, the 4-H program has, over time, worked through challenges when the approach to be inclusive of all hasn't felt as natural as it should. As outlined in the 4-H program history, there have been times when the 4-H program needed to shift its focus in order to allow for a more inclusive environment. The difficult reality is that the program sometimes faces volunteers who are not supportive of including all types of children or they don't feel equipped or educated enough to handle situations outside of what they are accustomed to. In order for 4-H programs to be high quality, they require staff members that are trained well, positive, and knowledgeable about their work. The challenge, however, is finding ways to support the professionals who create and deliver the programs to youth (Astroth, 2007).

As Extension moves forward, the 4-H movement must ask itself some tough questions. While the inclusion of all is core to our philosophical belief, are we truly postured to model this? The actions of some of our faculty, volunteers, and youth peers have demonstrated otherwise. The tough questions will be focused on acknowledging our shortcomings and being deliberate about how we move ourselves forward to openly discuss and address these topics. To truly "Make the Best Better," we must accept the fact that we must continue to improve in this area. We must be willing to do the hard work it will take to do so.

NOTES

1. Nonformal Education: Education that occurs outside of the formal school system.
2. Comparative Studies: comparing two or more things with a view to discovering something about one or all of the things being compared.
3. National Institute of Food and Agriculture (NIFA): A federal agency within the U.S. Department of Agriculture (USDA) that is part of USDA's Research, Education, and Economics (REE) mission area. The agency administers federal funding to address the agricultural issues impacting people's daily lives and the nation's future. (https://nifa.usda.gov/about-nifa).
   U.S. Department of Agriculture (USDA): The federal executive department responsible for developing and executing federal laws related to farming, forestry, rural economic development, and food.
4. Homogeneous: Consisting of parts of all the same thing.
5. Access: Freedom or ability to obtain or make sure of something.
   Equity: The quality of being fair and impartial.
6. Sexual Orientation: The scientifically accurate term for an individual's enduring physical, romantic, and/or emotional attraction to members of the same and/or opposite sex, including lesbian, gay, bisexual, and heterosexual (straight) orientations.
7. Electronic mailing from Dr. Muquarrab A. Qureshi, official at the National Institute of Food and Agriculture, March 28, 2017. Distributed nationwide to State 4-H program leaders via the National

Program Leaders Working Group (PLWG) and the Extension Committee on Organization and Policy (ECOP).

REFERENCES

American Academy of Pediatrics. (2018). Coming Out: Information for Parents of LGBTQ Teens. Healthy Children.org. https://www.healthychildren.org/English/ages-stages/teen/dating-sex/Pages/Four-Stages-of-Coming-Out.aspx.

Arnold, M. E., & Gagnon, R. J. (2019). Illuminating the Process of Youth Development: The Mediating Effect of Thriving on Youth Development Program Outcomes. *Journal of Human Sciences and Extension*, 7(3), 24–51.

———. (2020). Positive Youth Development Theory in Practice: An Update on the 4-H Thriving Model. *Journal of Youth Development*, 15(6), 1–23. doi:10.5195/jyd.2020.954.

Astroth, K. (2007). Making the Best Better: 4-H Staffing Patterns and Trends in the Largest Professional Network in the Nation. *Journal of Youth Development*, 2(2), 0702FA001.

Borden, L., Perkins, D., & Hawkey, K. (2014). 4-H Youth Development: The Past, the Present, and the Future. *Journal of Extension*, 52(4), 4COM1.

Copeland, A., Gillespie, T., James, A., Turner, L., & Williams, B. (2009). 4-H Youth Futures—College Within Reach. *Journal of Extension*, 47(1), 1IW1.

Crowder, C. (2018). Iowa 4-H Is Reaching Out to LGBTQ Kids, and It's Causing an Uproar among Conservatives. *Des Moines Register*, April 14. https://www.desmoinesregister.com/story/news/2018/04/13/4-h-lgbt-inclusion-guidelines-cause-uproar-iowa-john-paul-chaisson-cardenas-transgender-rights/513058002/.

Extension Committee on Organization and Policy. (2019). Personal correspondence issued via ECOP on behalf of ECOP Chair Jones, E, and National 4-H Council CEO, Sirangelo, J. (correspondence issued 2019).

Hensley, S., Place, N., Jordan, J., & Israel, G. (2007). Quality 4-H Youth Development Program: Belonging. *Journal of Extension*, 45(5), 5FEA8.

Hersher, Rebecca. (2017). Climate Scientists Watch Their Words, Hoping to Stave Off Funding Cuts. National Public Radio. https://www.npr.org/sections/thetwo-way/2017/11/29/564043596/climate-scientists-watch-their-words-hoping-to-stave-off-funding-cuts.

Howard, J. 2019. Developing Understandings of LGBTQ+ Communities for Volunteers to Enhance Acceptance and Inclusion in Extension Programming (abstract). National Extension Conference on Volunteerism. http://necv.msuextension.org/conf_agenda/workshops.html.

Human Dignity Trust. (2020). Map of Countries that Criminalise LGBT People. https://www.humandignitytrust.org/lgbt-the-law/map-of-criminalisation/.

Joelle, M., & Lofton, J. (1990). *Bittersweet Perspectives on Maryland's Extension Service.* Princess Anne: University of Maryland Eastern Shore.

Klemmer, C., & Zajicek, J. (2002). Growing Minds: The Effect of School Gardening Programs on the Science Achievement of Elementary Students. PhD Dissertation, Texas A&M University, College Station.

Lerner, R. M, Lerner, J. V., & Almerigi, J. B. (2005). Positive Youth Development, Participation in

Community Youth Development Programs, and Community Contributions of Fifth-Grade Adolescents: Findings from the First Wave of the 4-H Study of Positive Youth Development, *Journal of Early Adolescence*, 25(1), 17–71. doi:10.1177/0272431604272461.

Manor, A., & Pronovost, E. (2007). 4-H and Home Demonstration among African Americans: North Carolina's African American Extension Service. NCPEDIA, NCSU Libraries.

National 4-H Council. (2019). 4-H Youth Development. May. http://www.4-h.org/.

Peters, S. J., Alter, T. R., & Shaffer, T. J. (Eds.). (2018). *Jumping into Civic Life: Stories of Public Work from Extension Professionals*. Dayton, OH: Kettering Foundation Press.

Program Leaders Working Group. (2020). Practices for Inclusion of Individuals of All Genders and Sexual Orientations. Access, Equity and Belonging LGBTQ Support Site. https://access-equity-belonging.extension.org/wp-content/uploads/2020/02/Inclusion-Practices-Gender-and-Orientation-Final-2-20-2020.pdf.

Soule, K. (2017). Creating Inclusive Youth Programs for LGBTQ+ Communities. *Journal of Human Sciences and Extension*, 5(2), 103–125.

Van Horn, B. E., Flanagan, C. A., & Thomson, J. S. (1999). Changes and Challenges in 4-H (Part 2). *Journal of Extension*, 37(1), 1COM1.

Developing Social Capital as a Conduit to Social Justice

Matt Calvert, Keith Nathaniel, and Manami Brown

Cooperative Extension's signature youth development program, 4-H, was created over a century ago to prepare youth for an emerging agricultural and economic transformation, often casting youth as the agents of change within family and community economies. 4-H clubs have since provided venues for youth to "learn by doing," developing and practicing leadership skills, with the support of adults and more experienced youth. The structure of 4-H clubs, as public entities bringing youth and adults together from multiple families, and the orientation toward community engagement have built social capital for youth members, and then for the broader community, wherever 4-H is active. This chapter will focus on social capital as an outcome of the 4-H positive youth development approach and highlight practices that lead to broader community benefits and to opportunities to advance social justice.

What Is Social Capital?

Social capital is the connection among individuals—their social networks—and the norms of reciprocity and trust that develop from that connection (Chazdon et al., 2013; Putnam, 2000). Social capital benefits individuals who are privileged by access to networks and enjoy a sense of community and supportive relationships (Bourdieu, 1986; Putnam, 2000). Most relevant for Extension's work in community development, conceptualizing social capital as a factor in communities, for example, the extent to which a community has social relationships characterized by trust and reciprocity that benefit all members of a community, brings focus to both community and youth development. Chazdon, Allen, Horntvedt, and Scheffert of the University of Minnesota Center for Community Vitality created a community social capital model rooted in three distinct types of social networks: *bonding* networks among residents with a common social background, *bridging* networks among residents from different social backgrounds, and *linking* networks between residents and organizations and systems (Chazdon et al., 2013, p. 10).

Chazdon et al. (2013) place efficacy of community members believing they can make a difference at the center of their model of social capital, reflecting its importance as the energy needed to activate community social capital for collective action. Moving out from the center are two concentric circles with the inner ring focusing on trust and the outer ring focusing on engagement. "Trust" includes residents bridging networks by trusting others with different social backgrounds, linking networks that help them trust organizations and systems that can help them gain resources and bring about change, and bonding networks that help residents with a common social background to trust each other. Similarly, "engagement" helps residents to recognize the importance of bridging, linking, and bonding networks.

Extension programs have been targeted at developing all three types of social capital. Bonding networks are close ties that help people "get by" by offering a sense of identity and security (Catts & Ozga, 2005). These connections are usually with family, friends, and neighbors. Many 4-H clubs are places where youth build close and supportive relationships (Fields, 2016a; Enfield & Nathaniel, 2013), which are a key ingredient in positive youth development (Kress, 2005; Arnold, 2018).

Bridging networks are weaker ties that can help people get ahead and gain opportunities by widening the social network (e.g., to include community mentors and potential employers). Youth programs, such as 4-H, create opportunities for young people to build relationships with caring nonrelated adult volunteers and staff. Activities such as community service and youth leadership in community settings are especially powerful in building bridging networks for youth (Fields, 2016b). Lerner and Lerner (2013) have identified community contribution as a particularly significant youth development outcome of 4-H programs, as youth have opportunities within a community-based, public program. Finally, linking networks are links to organizations and systems that can help people gain resources and bring about change; in the Extension context, youth have opportunities to link to the resources of universities (Calvert, Emery & Kinsey, 2013). Positive youth development programs have many natural opportunities to strengthen social capital networks for youth.

Outcomes of youth programs are generally conceptualized at the individual level, with a focus on benefits to participants (Benson, Scales & Syvertsen, 2011; Lerner & Lerner, 2013) within community contexts (Gambone, Klem & Connell, 2002). However, programs that engage youth in communities can also have an effect on communities, as they create new trusting relationships. Over time, community organizations like Extension can promote community norms of cooperation and trust by engaging and linking both youth and adults in programs (Henness, Ball & Moncheski, 2013). Emery and Flora (2006) have identified one mechanism by which programs that begin with building bonding social capital with youth and adults can leverage those relationships into more significant ways of working together. A small community project can evolve into something transformative. Social capital development can serve as the community glue that facilitates other types of community development, for example, financial, human, and political capital (Flora & Flora, 2004). Intentional efforts to build social capital with youth are rarely at the center of youth development programs, but evidence is building that Extension's approach to youth and community development has built social capital throughout its history. However, not all communities have benefited from robust Extension programs, and the need to develop social capital is particularly significant for individuals and communities who are isolated from economic ties and other supportive networks (Wilson, 2009).

Social Capital as a Conduit to Social Justice

Fields and Nathaniel (2015) hypothesize that positive youth development programs can contribute to social capital and be used as a conduit toward social justice. There are programmatic strategies that, when implemented, help foster the development of social capital in 4-H youth development programs. Additionally, there are limitations with youth development models that fail to acknowledge the complex social, economic, and political forces that shape the lives of youth, particularly youth who have been historically marginalized (Ginwright & Cammarota, 2002; Fields, 2016a). Such forces include "issues of identity, racism, sexism, police brutality, and poverty that are supported by unjust economic policies" (Ginwright & Cammarota 2002, p. 82).

As outlined earlier, social capital represents linkages to people and resources that allow communities to solve problems together (Chazdon et al., 2013). Fields and Nathaniel (2015, para. 6) assert, "it is therefore plausible to consider the value of social capital in this context to 'solve problems' partly attributed to social injustice." In fact, "researchers have argued for years that people with high degrees of social capital—connection to community, networks, resources and institutions—are able to overcome barriers of social injustice" (Fields & Nathaniel, 2015, para. 6).

However, access to social capital, "is not equally available to all members of society of a given community" (Calvert, Emery & Kinsey, 2013, p. 5). In fact, some youth lack the social capital necessary to thrive through adulthood, especially if they grow up in families and neighborhoods that lack access to employers that might offer youth a first job, schools that link to broader opportunities, or community organizations that connect to decision-makers. The presence of social capital is a predictor of community action and engagement, which facilitates and enhances productive community development (Agnitsch, Flora & Ryan, 2009). Young people gain access to a multitude of opportunities, experiences, and forms of support in the areas of education, jobs and careers, emotional growth, and life skills development through connections formed at the family, school, and community levels. These connections and opportunities facilitate successful transition to adulthood and roles as productive change agents (Benson, 2003; Eccles & Gootman, 2002).

4-H Youth Development and Social Justice

The 4-H youth development program is arguably the largest and most well-known youth program in the country. Its history spans more than 110 years, through some of the United States' most tumultuous moments (i.e., Jim Crow, women's suffrage, the Great Depression) and significant moments (i.e., *Brown v. The Board of Education*, civil rights era). Through each of these movements and events, 4-H has stood as a pillar of youth and community connectedness through its vast networks of land-grant universities (created in 1862 and expanded to historically Black colleges and universities in 1890 and tribal colleges in 1994) supporting local educators and volunteers. Many families have participated in 4-H through multiple generations. However, these historical programs have not always included those most marginalized by social justice inequities. What is the result of youth programs that do not account for social justice? At worst, programs that reinforce unjust systems; at best, programs that do not meet the needs of many youth.

There are programs within 4-H that intentionally focus on developing the networks and efficacy among young people and communities that have been marginalized. The Children, Youth and Families at Risk (CYFAR) program is one such example where thirty-four states and a U.S. territory work with marginalized youth and families to strengthen communities, increase networks, and build efficacy. The 1890 and 1994 land-grant university 4-H programs have also demonstrated long-lasting community impacts through their work with youth and communities of color who may not have had access to such resources otherwise. It is critical for the greater 4-H community to learn from such examples.

The 4-H system has identified program approaches and practices that intentionally address issues of social justice through 4-H youth development (Fields & Nathaniel, 2015; Fields 2016a; Fields, Moncloa & Smith, 2018); these include the following:

- develop relationships with diverse caring adults who are culturally competent;
- increase trusting relationships—particularly bonding trust within families, friends, and neighbors;
- cultivate a genuine sense of belonging across different social networks—moving beyond inclusion and toward genuine belonging;
- engage in critical experiential learning and meaningful service-learning opportunities that address social issues and empower youth to take a critical stance against injustice;
- develop youths' sense of efficacy—the belief that they could make a difference in their communities; and
- foster linking connections between youth and institutions and organizations that can bring about change.

Case Study: Teen Corps

The Teen Corps Leadership program employs the best practices of youth and community development. It enables Extension to engage youth and adults in community development opportunities for the purpose of creating community change. Teen Corps was developed to meet a need in youth development programming in Baltimore, Maryland. Local demographic data, the Baltimore City Comprehensive Master Plan, and feedback from youth, adults, and community partners contributed to the development of the program.

Baltimore City has a population of approximately 622,000 and is comprised of diverse ethnic communities. With a growing exodus by individuals moving out of the city to surrounding counties, those that remain must find the means to revitalize and stabilize communities. Recent youth development research indicates there is a need to provide opportunities for youth that increase their leadership skills. These opportunities build on the assets of youth with the intent of strengthening their leadership capabilities. Engaging youth in leadership roles to work in partnership with communities captures the power of youth to affect community development, and engages communities to embrace their role in the development of its youth (Eccles & Gootman, 2002). Like many cities undergoing urban-renewal initiatives, it is imperative that Baltimore City Extension, 4-H youth/adult leaders, and community stakeholders assess community assets so that a clear strategy toward community revitalization can occur.

Program Delivery

Teen Corps is a citywide collaborative between 4-H youth and adult leaders, who partner to strengthen communities and increase leadership opportunities for youth ages twelve to eighteen. Teen Corps is a youth development program that helps youth reach their full potential.

A core element of the program is a monthly meeting that brings youth and adult representatives of 4-H clubs to one location that offers opportunities to receive leadership training. This format reinforces bonding social capital among members of each club and creates opportunities to build bridging relationships among youth and adults from different neighborhoods. Teen Corps members develop activities and projects that focus on service-learning, entrepreneurship/workforce readiness, and environmental science. Teen Corps youth and adult leaders learn together and build skills such as public speaking/communications, facilitation, community asset mapping, event planning, and training in service-learning, entrepreneurship/workforce readiness, and environmental science. They then take the skills learned back to their 4-H clubs and communities and help others learn and develop those same skills. After the youth/adult at the club sites learn the skills, Teen Corps members help them create projects and activities that are sustainable.

Teen Corps is an international collaborative through the development of a partnership with the Senegalese American Bilingual School (SABS) in Dakar, West Africa that was established in 2007–2008. Youth and adults from Senegal came to Teen Corps meetings in Baltimore City to provide an overview of Senegal and its culture and to learn about the Teen Corps model. The youth and adult leaders of their program engaged its student population in learning skills and implementing projects in leadership development, entrepreneurship, and environmental science.

Inputs

Teen Corps relies on a variety of financial resources, materials, supplies, and technology. First are community-based volunteers invested in youth development (bonding and bridging social capital), as well as Extension educators trained in 4-H youth development, community development, and applicable curricula. There are also collaborations in place with partnering agencies, community collaborators, schools, and after-school programs, as well as university resources (an example of linking social capital).

Outcomes

Participation in Teen Corps has various short-term, intermediate, and long-term outcomes. In the short term youth and adult participants complete the Teen Corps training, which increases their skills and knowledge in leadership, group processes, community asset mapping with GPS & GIS technology, service-learning and partnerships, life skills, entrepreneurship, and environmental science.

At the intermediate level, youth design, implement, and evaluate sustainable community-based activities such as youth-led business ventures, science, technology, engineering, and mathematics (STEM) and other greening-related projects, and workforce development programs. This results in an increased capacity of communities for economic resiliency through entrepreneurial, workforce development and educational learning opportunities, as well as increased participation in public issues education, community decision-making, and leadership development.

Long-term outcomes are the impact the Teen Corps program has on youth and adults. For example, youth could enter college or the workforce, especially in the areas of STEM, with skills that enable them to be successful; they might engage in entrepreneurial activities and/or start businesses that create economic opportunities for them and their communities; or they might engage in civic leadership activities that promote sustainable, resilient communities. On a community level, individuals and families fulfill basic human needs for safe, healthy, and adequate food, shelter, and environment; individuals, families, and groups build caring, safe, and productive communities; and individuals and families solve problems, make decisions, and embrace change to reach their full potential as individuals and as members of the community. To further these goals, high-quality educational programs, relevant and important to community members, are delivered by UME–Baltimore City employees.

A variety of methods have been used to identify and measure the impacts of this program. Through surveys, participants reported skills and knowledge increases. Program outreach impact was measured by the number of training sessions conducted by Teen Corps members. Additional outcomes are reflected in the number of educational and community development initiatives, focused on leadership, youth entrepreneurship, and environment, that have been led by the Teen Corps members.

Over the past sixteen years, Teen Corps has become an increasingly successful youth platform, reaching, engaging, and empowering over two hundred youth and adults who participated in Teen Corps for an average of three or more years. Teen Corps members have implemented citywide initiatives since 1998, in the areas of service-learning, entrepreneurship/workforce readiness, and environmental science, impacting over seventeen thousand Maryland residents. Teen Corps youth and adult leaders become positive role models in their communities, which lead to youth-led initiatives that engage communities in civic, business, community mapping, and other science-related projects. Examples of these projects and initiative outcomes:

- Implemented community mapping and service-learning projects (citywide) (social capital, built capital)
- Cofacilitated community-wide Master Plan meetings in the community of Upton to promote revitalization efforts (social capital)
- Presented Teen Corps at conferences, and has been featured on the Kellogg Foundation website (locally, statewide, and nationally) (human capital)
- Implemented the Baltimore City 4-H Youth Expo, a local event that showcases the talents of youth in competitive and noncompetitive projects and activities (sixteen years) (social capital, human capital)
- Received Train the Trainer Certificate of Completion (thirty hours) (human capital, social capital)
- Established entrepreneurial ventures (citywide)—(financial capital)
- Established greening projects (citywide)—(natural capital)

Teen Corps outcomes also extend beyond the programming with youth in Baltimore:

- The international exchange and partnership experience with a school in Dakar, Senegal, West Africa led to an entrepreneurship program in which Senegalese members implemented a school store selling snacks and other items to students, parents, and community members. The members also developed an environmental science project, the "SABS Goes Green Project," which gives an opportunity to change the habits and mentality of the students and their families.
- Teen Corps youth garnered additional leadership opportunities in civic engagement, entrepreneurship, and environmental science after aging out of 4-H.
- The Teen Corps Leadership curriculum was accepted through the UME Publication system and is a National 4-H Peer Reviewed curriculum available in English and Spanish.
- The Educator has secured over $200,000 in funds to support the local, state, national, and international programming implemented through the Teen Corps program.

Conclusion and Promising Practices

Teen Corps develops social capital for both individuals and the community. Regularly scheduled meetings develop bonding social capital by bringing youth and adult representatives of 4-H clubs to one location. Bridging networks are established when trained youth/adult leaders take the skills they've learned back to help their clubs and communities develop those same skills. Participants with different social backgrounds learn to trust and engage with each other. Finally, efficacy is developed by engaging in community development and sustaining long-term community projects.

When youth and adults work together, young people become more educated, independent, and responsible. This prepares young people with the skills and social capital to take their rightful place as caring, competent citizens and leaders in their community and the world. The Teen Corps Program integrates the Maryland 4-H youth development priorities: quality programs, youth/adult partnerships, outreach to underserved youth and adults, volunteer development, and strengthening 4-H clubs and groups.

Teen Corps is a model Extension outreach education program that addresses the need to expand/enhance program delivery and positive youth development in a variety of youth-centered educational settings that reach underserved and underrepresented youth.

REFERENCES

Agnitsch, K., Flora, J., & Ryan, V. (2009). Bonding and Bridging Social Capital: The Interactive Effects on Community Action. *Community Development*, 37(1), 36–51.

Arnold, M. E. (2018). From Context to Outcomes: A Thriving Model for 4-H Youth Development Programs. *Journal of Human Sciences and Extension*, 6(1), 141–160.

Benson, P. L. (2003) Developmental Assets and Asset-Building Community: Conceptual and Empirical Foundations. In R. M. Lerner and P. L. Benson (Eds.), *Developmental Assets and Asset-Building Communities* (pp. 19–43). Boston: Springer.

Benson, P. L., Scales, P. C., & Syvertsen, A. K. (2011). The Contribution of the Developmental Assets

Framework to Positive Youth Development Theory and Practice. *Advances in Child Development and Behavior*, 41, 197–230.

Bourdieu, P. (1986). The Forms of Capital. In J. Richardson (Ed.), *Handbook of Theory for Sociology of Education* (pp. 241–258). Westport, CT: Greenwood Press.

Calvert, M., Emery, M., & Kinsey, S. (2013). Issue Editors' Notes. In M. Calvert, M. Emery, & S. Kinsey (Eds.), *New Directions for Youth Development* (pp. 1–8). Hoboken, NJ: Wiley Periodicals.

Catts, R., & Ozga, J. (2005). *What Is Social Capital and How Might It Be Used in Scotland's Schools?* CES Briefings 36. Edinburgh: Centre for Educational Sociology, University of Edinburgh.

Chazdon, S., Allen, R. P., Horntvedt, J., & Scheffert, D.R. (2013). Reflecting (on) Social Capital: Development and Validation of a Community-Based Social Capital Assessment. Unpublished manuscript, University of Minnesota Extension.

Eccles, J., & Gootman, J. A. (2002). *Community Programs to Promote Youth Development.* Washington, DC: National Research Council.

Enfield, R., & Nathaniel, K. (2013). Social Capital: Its Constructs and Survey Development. In M. Calvert, M. Emery, & S. Kinsey (Eds.), *New Directions for Youth Development* (pp. 15–30). Hoboken, NJ: Wiley Periodicals.

Emery, M., & Flora. C. (2006). Spiraling-Up: Mapping Community Transformation with Community Capitals Framework. *Journal of the Community Development Society*, 37(1), 19–35.

Fields, N. (2016a). The Contribution of Urban 4-H Positive Youth Development towards Social Capital & Social Justice. PhD dissertation, Morgan State University, Baltimore.

———. (2016b). *Increasing Social Capital through Culturally Relevant Positive Youth Development.* Fact Sheet FS-1048, University of Maryland Extension.

Fields, N., Moncloa, F., & Smith, C. (2018). 4-H Social Justice Youth Development: A Guide for Youth Development Professionals. eXtension. https://dei.extension.org/2018/11/4-h-social-justice-youth-development-a-guide-for-youth-development-professionals/.

Fields, N., & Nathaniel, K. (2015). Our Role in and Responsibility toward Social Justice. *Journal of Extension*, 53(5), 5COM2.

Flora, C. B., & Flora, J. L. (2004). *Rural Communities: Legacy and Change.* Boulder, CO: Westview Press.

Gambone, M. A., Klem, A. M., & Connell, J. P. (2002). *Finding Out What Matters for Youth: Testing Key Links in a Community Action Framework for Youth Development.* Philadelphia: Youth Development Strategies.

Ginwright, S., & Cammarota, J. (2002). New Terrain in Youth Development: The Promise of a Social Justice Approach. *Social Justice*, 29(4), 82–95.

Henness, S., Ball, A., & Moncheski, M. (2013). A Community Development Approach to Service-Learning: Building Social Capital between Rural Youth and Adults. In M. Calvert, M. Emery, & S. Kinsey (Eds.), *New Directions for Youth Development* (pp. 75–95). Hoboken, NJ: Wiley Periodicals.

Kress, C. (2005). *Essential Elements of Positive Youth Development.* Washington DC: National 4-H Headquarters.

Lerner, R. M., & Lerner, J. V. (2013). *The Positive Development of Youth: Comprehensive Findings from the 4-H Study of Positive Youth Development.* Washington, DC: National 4-H Council.

Putnam, R. D. (2000). *Bowling Alone: The Collapse and Revival of American Community.* New York: Simon & Schuster.

Wilson, W. J. (2009). *More than Just Race: Being Black and Poor in the Inner City.* New York: Norton.

Addressing Diversity at Multiple Levels of the Social Ecological Model

Katherine E. Soule and Shannon Klisch

Cooperative Extension (CE) work is diverse and impacts a broad cross section of society and communities. CE agents are situated within communities, with the goals of bringing the research and evidence base of the university to the local communities, and bringing the lived experience and local knowledge of community members back to the university in communities that otherwise may not have access to university resources and engagement. Given this mission, CE is uniquely positioned to operate within a social ecological framework for improving community health and promoting equity. In this chapter, we begin by defining the social ecological model (SEM) and its application to increasing equity through CE work. We will go deeper into the details of each of the levels of the social ecological model and case studies regarding how CE can potentially or is currently working at different levels of the SEM to bring about more equitable improvements in community health. In some cases, CE may be doing the work without intentionally aligning each aspect to an evidence-based framework, such as the social ecological model. In other cases, CE may benefit from incorporating a SEM lens into existing programming and community assessments as we work toward meeting the needs of a changing and increasingly diverse society and enhancing equity in our program delivery.

Overview of the Social Ecological Model

The social ecological model is a framework used across many disciplines including human development, health education, psychology, and natural resource management.[1] The SEM has been used in program planning and evaluation and in assessing contributing factors to community and individual health outcomes. In relation to promoting racial and social equity, the SEM can be helpful as a tool to analyze the broader social and political contexts in which individuals make decisions rather than focusing solely on the individual as the agent of change. The SEM takes a broad view of human behavior, asserting that interrelated factors ascribed to different levels on the SEM are important influencers. The details of each level of the SEM vary depending on the context or discipline but generally include, and begin with, individual factors (genetics, knowledge, skills, an individual's experience of racism, privilege, etc.), and then move on to interpersonal (family life, social cohesion, access to professional networks and connections, microaggressions, etc.), organizational (quality of educational institutions, equitable hiring practices, organizational values), community (media portrayals, cultural and social norms, covert and overt racism), and public policy (written rules that govern access to public resources) environments in which individuals live, learn, work, or play. It can be helpful to think about SEM as overlapping rings that illustrate how factors at one level influence factors at another level.

The SEM is particularly useful when considering issues of equity and health in the context of CE programming because racial and social inequities occur at multiple levels of our social structure and educational institutions. Hurtado et al. (2012) propose a multicontextual model for diverse learning environments (DLE) in their work related to social justice in institutions of higher education. The DLE model builds upon substantial bodies of research related to the intersectionality of student, faculty, and staff identities and "the multiple contexts at work in influencing institutions of higher education and student outcomes for the twenty-first century" (Hurtado et al., 2012, p. 48).

CE generally places a high value on educating individuals and community members, the logic being that if people know better, they will do better. But what happens when they don't? Who is to blame when the health or agricultural outcomes are poor? Examining CE work through a SEM lens allows us to gain a broader picture including the environmental, social, and political factors that may facilitate or negate individuals' or communities' ability to act on their knowledge. In one study published in the *Journal of Extension*, researchers looked at factors within multiple levels of the SEM that affected seniors' ability to prepare and consume healthy food (Korlagunta et al., 2014). Using forty-nine independent variables hypothesized to impact older Supplemental Nutrition Assistance Program (SNAP) participants' ability to grocery shop and prepare food within the intrapersonal, interpersonal, and community levels of influence, researchers found significant correlations between self-reported abilities to shop and prepare healthy foods and social support, physical well-being, and food access—among many other factors examined across these levels of the SEM.

Situating CE work within the SEM allows for a broader level of understanding in relation to the conditions in which people live, learn, work, and play. Whereas previously the focus may have been on ensuring equality—that is, that all participants receive the same information at the same rate—the SEM allows the CE researcher or practitioner to ask deeper questions pertaining to equity: What would it take so that all individuals/communities had access to the resources needed to improve their lives and create positive changes in their communities? By looking through the perspective of a SEM framework, CE can step back from asking simply, "What do people need to know?" to "What resources are needed here? What political, social, or environmental conditions are impacting their ability to access those resources?"

Individual Level of Influence

At the individual level of influence, CE professionals can increase equity and inclusion while supporting changes in individuals' knowledge, attitudes, and behaviors that meet the needs of diverse clientele.[2] To do this, CE professionals should engage specific communities to

URBAN INTEGRATED PEST MANAGEMENT (IPM) PROGRAM

Although landlords in California are legally required to provide and maintain housing free of pests, such as bedbugs, tenants are sometimes held financially responsible. This practice is usually illegal, though low-income tenants often do not have knowledge of their rights or the fiscal means to hire legal representation in these matters. For undocumented individuals, there is also often a hesitancy to seek assistance from landlords. Recognizing the need for individuals to be able to effectively and safely manage bedbugs, academics in the Urban IPM Program developed research to better understand the needs of low-income, Spanish-speaking tenants (Campbell, Sutherland & Choe, 2016). These efforts led to the development of English- and Spanish-language educational materials to help low-income individuals prevent and eradicate these pests. For more information, see Romero et al. (2017).

understand their training needs and interests, and then developing trainings and research to meet those needs.

It is important for CE professionals to consider who is and who is not participating in their programs, trainings, research, and field days. CE professionals can do this by looking at the demographics of the population within their service area and comparing it to their program attendance, then asking difficult questions about who is not participating and why. In order to develop effective and relevant trainings, CE professionals need to conduct community assessments to ascertain existing community expertise, needs, and interests. What topics are relevant to this community? Is the cost to participate too high for these community members? What are the barriers preventing this community from participating? What knowledge already exists in this community and would be important to share more broadly? These questions likely can only be fully explored in partnership with community members. If CE professionals do not engage the community to learn about their lived experiences, it is likely that training topics, materials, and locations will be based off of CE professionals' assumptions about what a particular community needs, which can result in lower participation and ongoing unmet needs.

Interpersonal Level of Influence

At the interpersonal level of influence, CE professionals can increase equity and inclusion while supporting interpersonal processes, including working across peer, family, and groups that provide identity and support.[3] To do this, CE professionals should engage key community leaders and stakeholders in the specific communities that they would like to reach. These key individuals can help CE professionals understand the social structures and influences that shape the communities in which they hope to work, as well as to access social networks.

Working with key influencers may enable CE professionals to engage new communities in their efforts. Understanding the social structures and influences for a particular community may also create new avenues for CE programming and research to meet particular needs. For example, understanding what produce a community prefers, CE professionals working in agriculture may be

FARM ADVISORS WORKING WITH PUNJABI COMMUNITIES

CE professionals working in counties with large Punjabi populations found that Punjabi farmers were a close community that often shared farming practices and information within their community (as opposed to looking to attending educational events hosted by outside experts). CE professionals built a partnership with the local Punjabi radio show host, who was well respected and largely listened to. The radio host often invites CE professionals onto his show to share information about research and education that impacts Punjabi farming practices in the area. As trust was built between CE professionals and the Punjabi farming communities, farmers would often request and value one-on-one consultations with CE professionals. For more information, see Cantor and Kresge (2009).

able to support small farms in growing and selling those crops within their community. Learning about who makes the financial management decisions in families may increase CE professionals' ability to develop and deliver relevant financial literacy programs. Finding out where communities gather can help CE professionals identify potential programming sites for trainings and field days that would be welcoming and convenient for participants.

Organizational Level of Influence

At the organizational level of influence, CE professionals can increase equity and inclusion while ensuring inclusive policies, practices, and processes, including hiring practices, allocation of resources, and targeted program sites.[4] To do this, CE professionals should engage appropriate administrative and leadership personnel in the review and development of policies, practices, and processes that they would like to address. At the same time, CE professionals often have agency at the local level to address organizational factors that can impact inclusion, for example,

MASTER GARDENER PROGRAM WORKING WITH INCARCERATED YOUTH

Research indicates that gaining life skills and marketable technical experience supports healthy behavior postrelease, explaining the lower recidivism rates among those who have been involved with prison gardening programs (Jenkins, 2016). CE personnel in California identified the juvenile hall as a targeted program site to work with incarcerated youth through the Master Gardener Program. Partnering with the juvenile hall, volunteers and staff taught incarcerated youth and adult officers how to build and maintain a garden, and plant and harvest vegetables at the detention center. The youth also learned how to prepare produce, which was then made into meals, a new skill for many of the participants. Produce from the garden was also donated to the local food pantry. The juvenile hall manager shared that "the kids were very aware and proud of the fact that the food went to a pantry in the county, [that] they could provide for someone else" (Gillison & Reiter, 2017).

selecting program sites to target specific populations and reduce barriers, developing materials in appropriate languages, or recruiting/hiring staff from populations to be served.

Cooperative Extension has a significant opportunity at the organizational level to address equity, inclusion, and diversity. We should be evaluating our recruitment and hiring decisions to monitor how and when implicit bias impacts search committee recommendations. Cooperative Extension can and should monitor how we train search committees to be aware of and check bias, how implicit bias impacts salary analysis, and what professional development opportunities around inclusion and equity are being offered to current CE professionals. Further questions that should be asked are the following: How can we better resolve communication differences to support inclusive work environments? When looking at our distribution of resources, programs, and offices, we should be aware of who is and is not being served. Can we reallocate our resources to better serve all in our communities? How can we address the historic disparities in access to CE programs? Which industry and community-based organizations are we partnering with? Who should we partner with to reach communities that have been excluded from CE programming? To make positive impacts in these areas, we need to intentionally evaluate and address how our policies, practices, and procedures support (or do not support) the development of an inclusive society. We may need support and expertise beyond our own organization to be effective in these efforts, particularly when first learning about how such organizational decisions can be inclusive to dominant populations while marginalizing diverse populations.

Community Level of Influence

At the community level of influence, CE professionals can increase equity and inclusion while facilitating individual behavior change by supporting community-level resources, communication, and positive norms.[5] To do this, CE professionals should engage with appropriate community-based organizations, coalitions, and media to assess the educational needs and behavioral changes they would like to address.

While working to create community-level change can impact large populations and address structural barriers and inequities, these efforts also can have large negative impacts if not conducted cautiously. In fact, misaligned community-level change efforts can increase inequities, barriers, and conflicts. To be effective and have positive results, CE professionals should ensure that they have meaningful engagement with all stakeholders. In particular, we need to be aware of differences in cultural norms, especially around communication, for the populations we are partnering with. If we are not mindful of communication norms we might miss differences in nonverbal and verbal communications and mistakenly believe we have agreements when our partners are actually expressing concern. We need to be aware of our true capacity and resources when planning projects to avoid overpromising and leaving communities with unfulfilled commitments. Once programs have been implemented, we need to work with our partnering communities to evaluate programs for effectiveness and consistency in application. If something is not working as anticipated, we need to be prepared to adapt our work to ensure the impact is positive and beneficial from the perspective of our community partners.

REVITALIZING NATIVE AMERICAN FOOD AND PLANT CULTURE

CE academics in California have built a decade-long relationship with three Native American communities. This relationship has focused on community-driven research, education, and outreach to improve tribal health and food security among the tribal communities, and enhance the quality and availability of culturally significant plants. Fifteen tribal staff have been employed, providing hands-on educational opportunities to manage, grow, gather, prepare, and preserve local and traditional foods. Nearly seven thousand tribal people have participated in the activities, including youth who learned how to analyze scientific data (Kim et al., 2019). In total, tribal and academic partners have secured $5.2 million from the U.S. Department of Agriculture (USDA)-Agriculture and Food Research Initiative to strengthen tribal capacity for ongoing food security and food sovereignty work, and to expand field research and digital data analysis to enhance the resilience of Native American foodways under changing climate conditions.

To help expand these efforts across the state, a group of CE professionals led workshops for CE colleagues on how to create successful working relationships with tribes. These successes have led to the development of a newly formed Native American Communities Partnership workgroup focusing on strengthening CE's capacity and partnerships with Native American communities in California. For more information on the collaboration, projects, and impacts, see Sowerwine et al. (2019a); Sowerwine et al. (2019b); and https://nature.berkeley.edu/karuk-collaborative/.

Public Policy Level of Influence

At the policy level of influence, CE professionals can increase equity and inclusion when working to improve or implement policies with a focus on eliminating disparities.[6] To do this, CE professionals should engage with the public and policymakers to provide science-based research information to support the development, improvement, or implementation of policies.

Many local, state, and federal agencies support policies and laws that are intended to increase equity and eliminate disparities. CE professionals can support community comprehension and implementation of the policy decisions, as well as ensuring that information is accessible for a wide range of audiences. CE researchers may also explore the consequences of policy decisions on varying populations, as well as the impact of differences in policy implementation. At a local level, CE professionals can provide additional guidance to explain how policies that impact diversity and inclusion can be supported by our organizations and communities.

Using the SEM to Promote Equity in Nutrition Programming

In this section, we consider how CE might work across the social ecological model's levels of influence in a single program area to expand our outcome and impacts. This version of the SEM is utilized by CE programs funded through the USDA's SNAP-Ed nutrition education program and the 2015–2020 Dietary Guidelines for America. CE practitioners can use the model to plan programs that address the various individual factors, settings, sectors, and social and cultural

NUTRITION POLICY INSTITUTE

The Nutrition Policy Institute (NPI) works to improve federal nutrition programs and policies, pursue structural changes in food systems, and improve physical and social environments in order to improve nutrition and prevent obesity and chronic diseases. This work is accomplished by CE academics who conduct and evaluate the impact of programs and policies, while communicating scientific information to policymakers and the public. NPI focuses on health disparities and community engagement, serving as a resource for federal, state, and local legislators. For example, NPI conducted surveys to evaluate the impact of California's Healthy Beverages in Child Care Act (AB 2084). Although they found evidence that the law was effective in improving the beverage environment of childcare settings, they also found only one quarter of participants in the state were in full compliance. As a result, they developed and piloted an online training for childcare providers to increase comprehension and compliance. For more information, visit NPI at https://npi.ucanr.edu (2019).

norms that impact a person's nutrition and physical activity behaviors and, ultimately, their health outcomes.

Individual factors that may impact health outcomes include a person's age, sex, socioeconomic status, race or ethnicity, ability or disability, knowledge and skills, genetics, and food preferences. Historically SNAP-Ed-funded programs were focused almost exclusively on improving individual factors such as knowledge and skills related to healthy eating and physical activity. In recent years, however, there has been increasing recognition of the importance of a person's context and environment in relation to their ability to achieve healthy outcomes. This concept can be illustrated by thinking about a participant in a nutrition class who is learning how to prepare a healthy meal using fresh fruits and vegetables. The participant lives in a neighborhood where there are no full-service grocery stores, and the nearest corner market only carries processed canned foods and a few apples or bananas. This participant's ability to put their new knowledge and skills into practice is severely restricted by their access to these healthy foods. While education and skill building are crucial steps in supporting healthy individuals and communities, they are not sufficient.

The settings level of the SEM recognizes that individuals make health decisions in a variety of locations including where they live, learn, work, and play and that these environments or settings have dramatic impacts on health outcomes and life expectancy. A series of maps displaying life expectancy by zip code (VCU, 2016) shows large discrepancies between neighborhood life expectancy. In several cities, life expectancy can vary by up to twenty years between neighborhoods just five miles away from each other. This demonstrates the need for an examination of the environments in which individuals make health decisions. In CE programming, an examination of the environmental setting is inherent in the work of farm and agricultural advisors who may be monitoring and recording data related to the quality of the soil, air temperature, moisture, presence of beneficial insects and pests, etc. Similarly, humans are situated in and impacted by their environments whether they are a farmer that is trying to cut back on water use or a community member that is wanting to live a long and healthy life. Settings can enhance or inhibit an individual's access to the choices that are necessary to achieve health, and ignoring the settings in which people live can further perpetuate inequity by blaming the individual for choices that are out of their control (Fane & Ward, 2016). Further, the 2008 report from the World Health Organization Commission on the Social Determinants of Health outlined actions needed in order to substantially reduce global health inequities. In action 3, the commission states that an understanding of the social determinants of health by the general public and health practitioners is crucial to the elimination or reduction of health disparities (CSDH, 2008, p. 189). CE practitioners and researchers should intentionally engage learners in a critical understanding of the social determinants of health.[7] However, there is a lack of curricula available to specifically teach the social determinants of health and the importance of settings on health behaviors to diverse and community-based audiences.

Sectors that can impact individual behavior include government policies, business practices, educational policies, and health care systems among many others. Within CE, there continues to be a movement to ensure equitable use of government resources and access to services. This has been particularly important in the case of the 4-H youth development programs that have historically served mainly white clientele. Equal opportunity policies are part of the effort to expand access to government and university resources to nonwhite clientele so that communities can benefit equally. In the context of nutrition education programming, school wellness policies can lay the foundation for schools to promote health and nutritional equity. CE professionals can serve as technical assistance professionals to school wellness committees in order to ensure that policies reflect the best available research on wellness and are implemented with the support and engagement of the parents and students in the school district.

Social and cultural norms are both impacted by and have an impact on individuals, settings, and sectors. Social and cultural norms are the "shared assumptions of appropriate behaviors" (HHS & USDA, 2015, p. 66) based on the values of a community and a society. These social norms can influence food preferences, health and wellness priorities, and body image and permeate every aspect of our lives. CE professionals' work must understand and acknowledge the role of community-held social and cultural norms and plan programming, education, and interventions that are informed by these—or risk being irrelevant to the communities we are trying to reach.

Conclusion

While the social ecological model can be a useful framework for program planning to address diversity, equity, and inclusion in Cooperative Extension, we also need to be aware of our individual and organizational readiness to effectively conduct this work. Indeed, effectively working at different levels of the SEM requires ongoing training and support. Just as program supervisors train staff to implement evidence-based curricula, CE staff also need training to engage community members in work to change the policies, systems, and environments that impact health. Currently, trainings and curricula that build CE staff and community capacity to recognize the importance of the social determinants of health are lacking.

We should begin within, utilizing a self-reflexive and multidimensional approach to assess our readiness as individuals and as an institution. As individuals, our own beliefs and practices affect our ability to advocate the development of programs that support inclusiveness. We need to be aware of our own assumptions about diverse needs and perspectives, particularly questioning how we might minimize the intersectionality of individuals' lived experiences. A first step in the process is attending trainings on racial and gender equity and being open to learning about our own biases. Taking one or more of the Harvard implicit association tests (https://implicit.harvard.edu/implicit/) can bring to light prejudices that we hold as individuals, so we can begin to examine and, ultimately, deconstruct those prejudices. At an organizational level, decision makers within CE can promote trainings on racial and gender equity and provide resources for CE researchers and staff to attend and develop plans for improving their knowledge and practice when it comes to equity and inclusion. Further, these trainings on equity should be made available to community partners and to the public. A deeper understanding of health equity and the social determinants of health by CE researchers and our clientele is essential to reducing health disparities and promoting a more equitable society. As discussed in-depth by Hurtado et al. (2012, p. 104), "Intentional education with the aim of fostering civic equality reflects a belief that our students represent our best investment for a more just, equitable, economically viable, and stable democratic society."

How do our own social contexts impact our understanding of diversity? How do the ways we currently distribute our time and program monies support or hamper these efforts? How do our trainings and curricula focus on one aspect of the SEM and ignore the social determinants of health? Are we afraid to move this work forward and make mistakes? It may be that we as CE professionals and organizations need to increase our own intercultural competence and self-awareness as the first step in utilizing the SEM to create positive change.

NOTES

1. The social ecological model (SEM): A framework in assessing contributing factors and planning approaches to community and individual health outcomes, which analyzes broader social, political, and organizational contexts in which individuals, groups, and families make decisions rather than focusing solely on the individual as the agent of change.
2. Individual Level of Influence: Individuals' knowledge, attitudes, and behaviors.
3. Interpersonal Level of Influence: Interpersonal group (including families, peers, social, and work groups) social structures and influences.
4. Organizational Level of Influence: Organizational policies, procedures, practices, systems, and environments.
5. Community Level of Influence: Community level resources, communication, and norms.
6. Public Policy Level of Influence: Laws and policies, including development, implementation, and enforcement.
7. Social Determinants of Health: Environments and conditions that impact quality of life and health outcomes. Examples of social determinants of health include access to quality education, safe and affordable housing, availability of healthy and affordable foods that are culturally relevant, living and

working conditions that are free from environmental pollutants and toxins, and accessible and safe spaces for physical activity, to name a few (ODPHP, 2020).

REFERENCES

Campbell, K., Sutherland, A., & Choe, D.-H. (2016). UC Survey: When It Comes to Bed Bugs, Know What's Happening in Your Units. *California Apartment Management*, Spring, 18–19.

Cantor, A., & Kresge, L. (2009). Evaluation of the Environmentally Responsible Management Practices for Tree Crops in the Feather River Basin Project. California Institute for Rural Studies.

CSDH (Commission on Social Determinants of Health). (2008). *Closing the Gap in a Generation: Health Equity through Action on the Social Determinants of Health; Final Report of the Commission on Social Determinants of Health.* Geneva: World Health Organization.

Fane, J., & Ward, P. (2016) How Can We Increase Children's Understanding of the Social Determinants of Health? Why Charitable Drives in Schools Reinforce Individualism, Responsibilisation and Inequity. *Critical Public Health*, 26(2), 221–229. doi:10.1080/09581596.2014.935703.

Gillison, S., & Reiter, M. (2017). UC Master Gardeners Pilot Gardening Program for Incarcerated Youth. UC Agriculture and Natural Resources. https://ucanr.edu/sites/ucanr2/?impact=1082&a=0.

HHS (U.S. Department of Health and Human Services) & USDA (U.S. Department of Agriculture). (2015). *2015–2020 Dietary Guidelines for Americans.* 8th ed. December. http://health.gov/dietaryguidelines/2015/guidelines/.

Hurtado, S., Alvarez, C. L., Guillermo-Wann, C., Cuellar, M., & Arellano, L. (2012) A Model for Diverse Learning Environments. In J. Smart and M. Paulsen (Eds.), *Higher Education: Handbook of Theory and Research* (Vol. 27, pp. 41–122). New York: Springer.

Jenkins, R. (2016). Landscaping in Lockup: The Effects of Gardening Programs on Prison Inmates. *Graduate Theses & Dissertations.* Paper 6. https://scholarworks.arcadia.edu/cgi/viewcontent.cgi?article=1005&context=grad_etd.

Kim, K., Ngo, V., Gilkison, G., Hillman, L., & Sowerwine, L. (2019). Native American Youth Citizen Scientists Uncovering Community Health and Food Security Priorities. *Health Promotion Practice*, 21(1), 80–90.

Korlagunta, K., Hermann, J., Parker, S., & Payton, M. (2014). Factors within Multiple Socio-Ecological Model Levels of Influence Affecting Older SNAP Participants' Ability to Grocery Shop and Prepare Food. *Journal of Extension*, 52(1), 1RB3.

NPI (Nutrition Policy Institute). (2019). About Us. http://npi.ucanr.edu/About_Us/.

ODPHP (Office of Disease Prevention and Health Promotion). (2020). Social Determinants of Health. Healthy People. https://www.healthypeople.gov/2020/topics-objectives/topic/social-determinants-of-health.

Romero, A., Sutherland, A. M., Gouge, D. H., Spafford, H., Nair, S., Lewis, V., Choe, D.-H., Li, S., & Young, D. (2017). Pest Management Strategies for Bed Bugs in Multi-unit Housing: A Literature Review on Field Studies. *Journal of Integrated Pest Management*, 8(1), 13; 1–10. https://academic.oup.com/jipm/article-lookup/doi/10.1093/jipm/pmx009.

Sowerwine, J., Mucioki, M., Friedman, E., Hillman, L., & Sarna-Wojcicki, D. (2019a). *Food Security Assessment of Native American Communities in the Klamath Basin with the Karuk Tribe, Klamath*

Tribes, Yurok Tribe, and Hoopa Tribe. Karuk-UC Berkeley Collaborative. Berkeley, CA: University of California at Berkeley.

Sowerwine, J., Mucioki, M., Sarna-Wojcicki, D., & Hillman, L. (2019b). Reframing Food Security by and for Native American Communities: A Case Study among Tribes in the Klamath River Basin of Oregon and California. *Food Security,* 11, 579–607. doi:10.1007/s12571-019-00925-y.

VCU (Virginia Commonwealth University). (2016). Mapping Life Expectancy. Center on Society and Health. September 26. https://societyhealth.vcu.edu/work/the-projects/mapping-life-expectancy.html.

An Introduction to Environmental Justice and Extension Programs Engaging Vulnerable Communities

Sacoby Wilson, Helen Cheng, Davin Holen, Erin Ling, Daphne Pee, and Andrew Lazur

For over forty years, environmental justice (EJ) researchers have demonstrated that many low-wealth populations, populations of color, and marginalized and disenfranchised groups live in communities that experience a disproportionate risk from the burden of and exposure to environmental hazards. These hazards can include noxious land uses such as landfills, incinerators, publicly owned treatment works (e.g., sewer and water treatment plants), industrial animal operations, Superfund sites, Toxics Release Inventory facilities, power plants, chemical plants, heavily trafficked roadways, and other locally unwanted land uses and associated access to safe drinking water (Wilson, 2009, 2010; Bullard, 1994; UCC, 1987; Wing, Cole & Grant, 2000; Bullard et al., 2007; Bryant & Mohai, 1992). Due to the spatial concentration of environmental hazards, pollution-intensive facilities, and noxious land uses, the cumulative impact of environmental injustice leads to increases in negative health outcomes and community stress, and reduced quality of life and community sustainability (Wilson, 2009, 2010).

Additionally, communities impacted by environmental injustice may have differential access to health-promoting infrastructure, including grocery stores, parks, green space, safe housing, and other salutogenic resources (Wilson, 2009). Access to safe drinking water is a primary example of an environmental health disparity issue in environmental justice communities. There is increasing recognition of significant failures in the safety of public, regulated drinking-water supplies in the United States, including those in Corpus Christi, Texas; Flint, Michigan; and Charleston, West Virginia (Katner et al., 2016). Low-income populations, populations of color, and other disenfranchised groups are often disproportionately impacted by these crises (Schaider et al., 2019). After examining thousands of utilities over four years, Switzer and Teodoro (2017) reported that communities with higher populations of African American and Hispanic populations were significantly more likely to have both health-related and management violations of the Safe Drinking Water Act, and that these effects were exacerbated by poverty. Beyond communities with failing infrastructure, studies have also shown the lack of access to publicly regulated sewer and water infrastructure in low-wealth communities of color, including unincorporated communities, particularly in the South (Stillo & Gibson 2018; Leker & Gibson, 2018; Heaney et al., 2011; Wilson et al., 2008a, 2008b).

Many communities impacted by environmental justice issues face further risk from climate change. Climate change will both reveal and magnify environmental injustice because certain groups of Americans are disproportionately affected and less able to adapt to or recover from climate change impacts due to their race/ethnicity, geographic location, income, or status (Wilson et al., 2010; EPA, 2016). For instance, the intersection of race, class, geography, immigrant status, and health care access was quite visible in the pre-event and postevent aftermath of Hurricane Harvey in Houston, TX, in 2018 (Bodenreider et al., 2019). Many of the impacts observed in Houston are commonly experienced by communities with EJ issues who already face:

- greater vulnerability to climate change because residents disproportionately live in areas with more climate-related impacts such as flooding, severe storms, hurricanes, heat waves, and droughts;
- disproportionate impacts from underlying health disparities that can be exacerbated by climate change;
- limited or no access to health care service, which will be further reduced during or after climate events;
- economically unstable situations that hinder preparations for climate disasters and result in less ability to relocate or rebuild after a disaster, an effect known as "climate gentrification" (Keenan, Hill & Gumber, 2018; Anguelovski et al., 2019).

Defining Environmental Justice

The environmental justice movement is focused on redefining the environment as where we live, work, play, pray, and learn, and dismantling environmental racism and other processes such as segregation and inequities in housing, transportation, zoning, food access, access to natural amenities, and economic opportunity that drive environmental injustice. Bunyan Bryant, a pioneering EJ researcher, provides an expansive definition of environmental justice:

> Environmental justice is served when people can realize their highest potential, without experiencing the "isms." EJ is supported by decent paying and safe jobs, quality schools and recreation, decent housing and adequate health care, democratic decision-making and personal empowerment; and communities free of violence, drugs and poverty. These are communities where both cultural and biological diversity are respected and highly revered and where distributed justice prevails. (Bryant, 1995, p. 6)

The official definition of environmental justice from the U.S. Environmental Protection Agency (EPA) is more narrowly focused than the environmental justice principles developed through Bryant's definition:

> Environmental Justice is the fair treatment and meaningful involvement of all people regardless of race, color, national origin, or income with respect to the development, implementation, and enforcement of

> environmental laws, regulations, and policies. EPA has this goal for all communities and persons across this Nation. It will be achieved when everyone enjoys the same degree of protection from environmental and health hazards and equal access to the decision-making process to have a healthy environment in which to live, learn, and work. (EPA/OEJ, 2013)

The EPA's definition is problematic because it does not provide a framework for addressing the historic and cumulative burden of environmental hazards on communities of color, low-wealth populations, and indigenous peoples in the United States (Wilson, 2010). The definition also ignores both the social determinants of health as well as the historic and contemporary role of structural and institutional racism as a primary driver of environmental injustice (Wilson, 2010). The EPA also roots environmental justice in the concept of environmental equity, meaning all groups should share an equal burden of negative impacts, which is in opposition to the conservation and sustainability goals of the environmental justice movement (Wilson, 2010). Lastly, the definition does not include political empowerment or social justice, two important tenets of the environmental justice movement (Wilson, 2010). The grassroots environmental justice movement defined environmental justice for itself in 1991 through a meeting known as the First National People of Color Environmental Leadership Summit. At this summit, the seventeen Principles of Environmental Justice were written that cover topics including conservation, human rights, sustainability, war, chemical production, indigenous rights, workers' rights, and reparations (Wilson 2010).

Promising Practices for Environmental Justice and Community Engagement

Community-based EJ organizations have effectively utilized community-driven research (Corburn, 2005; Heaney et al., 2007; Heaney et al., 2011; Israel et al., 2005a, 2005b; Minkler & Wallerstein, 2003; Wilson et al., 2008a, 2008b; Wing, 2002; Northridge et al., 1999) to address local environmental justice and health issues, including air pollution in metropolitan areas, housing stock, basic amenities, pollution-intensive industries, access to healthy and affordable foods, locally unwanted land uses, transportation issues, and animal production. Through community-driven research, community groups utilize their grassroots activism and resources, expert local knowledge, and university partners to develop a framework to address and solve environmental health and justice issues at the local level.

This approach relies on community-based organizations getting more involved in creating a scientific agenda, thereby giving the neutral perspective found in conventional research an intentionality that aims to increase science's utility and equity. The contextual expertise of community members is thus valued, as are their lived experiences and their ability to employ a research empowerment framework. The community-driven approach allows for the research process to

- be more authentic and action-oriented;
- build and sustain a community's capacity to address environmental justice and health issues;
- increase civic engagement by minority and low-income stakeholders;

- bring equality to the relationships between local experts and academic experts;
- ensure that the research is locally relevant;
- develop an intergenerational pipeline of community leaders knowledgeable about the EJ issues; and,
- produce end-products that lead to justice, action, positive environmental and social change, solutions, and reduction of hazards, health risks, and environmental health disparities (Wilson et al., 2008a, 2008b; Heaney et al., 2011; Heaney et al., 2007; Corburn, 2005; Israel et al., 2005a, 2005b; Wing, 2002).

The use of community-driven research methods has helped empower communities, raise awareness on environmental justice issues at the local, state, and national levels, increase environmental health literacy, and enhance capacity-building and training.

Community-Based Participatory Research

To define community-driven research, we use the definition of community-based participatory research (CBPR) given by Israel et al. as:

> a collaborative approach to research that equitably involves, for example, community members, organizational representatives, and researchers in all aspects of the research process. The partners contribute "unique strengths and shared responsibilities" . . . to enhance understanding of a given phenomenon and the social and cultural dynamics of the community, and integrate the knowledge gained with action to improve the health and well-being of community members. (1998, p. 177)

This method of research is widely used in public health. It goes by many names and variants, but the basic principles are the same, identified by Israel et al. (1998, pp. 178–180) as:

- Recognizing the community as a unit of identity.
- Building on strengths and resources within the community.
- Facilitating collaborative partnerships in all phases of the research.
- Integrating knowledge and action for mutual benefit of all partners.
- Promoting a colearning and empowering process that attends to social inequalities.
- Involving a cyclical and iterative process.
- Addressing health from both positive and ecological perspectives.
- Disseminating findings and knowledge gained to all partners.

Establishing partnerships and using participatory research methods are central to CBPR. The Community-Campus Partnerships for Health (CCPH), a nonprofit devoted to forging community-academic partnerships to solve health issues, has developed elements of an "authentic partnership" that drive successful implementation of CBPR:

- Principles of mutual accountability, open communication, shared goals, and transparency.
- Quality processes to ensure healthy partnerships.
- Meaningful outcomes for communities.
- Transformative processes at the individual, community, institutional, and political level. (CCPH, n.d.)

CBPR and the Community Engagement Continuum

Because of the time and effort it takes for partnerships to successfully develop, CBPR is an evolutionary process. To reflect this, the Clinical and Translational Science Awards Consortium (2011) places community-university research partnerships on a community engagement continuum of which CBPR is just one component (see table 1). On this continuum, a fully realized CBPR project would fall under "Collaborate," as it engages communities in the research process and strengthens partnerships. The community engagement continuum shows that partnerships may vary from little interaction among communities and researchers to partnerships characterized by shared power, bidirectional communication, and trust.

Integrating Authentic Community Engagement into Extension Programs

Extension programs across the nation have long educated and engaged communities on environmental issues. While the authors of this chapter are unaware of any programs that specifically apply EJ principles to their efforts, many Extension educators have implemented programs that reflect the various phases of the community engagement continuum. In this section, we present three case studies of Extension efforts that span the community engagement continuum in engaging communities with EJ issues.

Extension in Action: Reaching Out to Homeowners to Ensure Safe Private Drinking Water in Virginia

Approximately 15 percent of the U.S. population relies on private water supplies that are unregulated by the Safe Drinking Water Act (SDWA) (Maupin et al., 2014). These 44.5 million people are responsible for routine testing and interpretation of water quality results, care and maintenance of the well or spring, and addressing problems through water treatment or other measures. Craun et al. (2010) state that although cases of waterborne illness have decreased overall in the United States since the 1980s, incidences of waterborne illness from private water supplies have increased. In a North Carolina study, Gibson and Pieper (2017) found that 99 percent of hospital visits for acute gastrointestinal illness from exposure to waterborne microbial contamination may be attributed to private wells.

One-fifth of Virginia's population (1.7 million people) relies on private water supplies (Maupin et al., 2014). Lower-income and less-empowered populations are more likely to be using older or poorly constructed or maintained wells, potentially in more densely populated areas or with little

Table 1. The Community Engagement Continuum

OUTREACH	CONSULT	INVOLVE	COLLABORATE	SHARED LEADERSHIP
Some community involvement	More community involvement	Better community involvement	Community involvement	Strong bidirectional relationship
Communication flows from one to the other, to inform Provides community with information	Communication flows to the community and then back, answer seeking Gets information or feedback from the community	Communication flows both ways, participatory form of communication Involves more participation with community on issues	Communication flow is bidirectional Forms partnerships with community on each aspect of the project from development to solution	Final decision-making is at community level
Entities coexist	Entities share information	Entities cooperate with each other	Entities form bidirectional communication channels	Entities have formed strong partnership structures
Outcomes: Optimally, establishes communication channels and channels for outreach	Outcomes: Develops connections	Outcomes: Visibility of partnership established with increasing cooperation	Outcomes: Partnership building, trust building	Outcomes: Broader health outcomes affecting broader community Strong bidirectional trust built

to no control over nearby land use. The Virginia Household Water Quality Program (VAHWQP) is a Virginia Cooperative Extension effort supported by grant funding and cost-recovery that analyses drinking water samples and educates private water supply users in at least sixty-five counties annually (Benham et al., 2016). Grant funding is often available to subsidize the cost of sample analysis ($60 per sample kit in 2019) if needed. Since 2008, about sixteen thousand samples have been analyzed for fourteen bacterial and chemical constituents.

The vast majority of VAHWQP samples (55–60 percent annually) do not meet the SDWA's health-based standards for total coliform bacteria, E. coli bacteria, sodium, lead, and copper (Pieper et al., 2015; Virginia Cooperative Extension, 2019). Evaluation surveys conducted immediately following VAHWQP drinking water clinics indicate that 93 percent of participants report they understand their results, and many plan to take action as a result of what they learn. A follow-up phone survey conducted in 2013 indicated that participants reported taking recommended actions at significantly higher rates in the two years following participation than stated immediately following the clinics (Benham et al., 2016).

While these results indicate that VAHWQP successfully educates and motivates homeowners to test their drinking water and repair or maintain their drinking water systems, we recognize there are greater challenges and barriers to action for lower-income families. Like many Extension efforts, this program was designed for a broader audience of homeowners, not just low-income households. While the cost of the water analysis can be subsidized relatively easily for our program, those living in poverty still face significant challenges with the time required to participate in programming and the costs required to address identified issues, which may include installing or maintaining water treatment devices and/or repairing or replacing a well. To help lower-income families address such issues, VAHWQP partner organizations, like the Southeast Rural Community Assistance Project, provide technical assistance and low-interest loans and grants to help participants repair or replace water systems. In many cases, even with this assistance, families still struggle to meet the requirements for the loans and grants or, because they rent their homes, are ineligible to receive funding or technical support.

While VAHWQP faculty have not yet identified how to address these barriers for low-income families, we continually look for opportunities to improve our impact and to better serve all individuals who rely on private water supplies. In 2015, VAHWQP adapted our programming model to begin working with rural schools and reach families who do not have the time, money, or perhaps trust to attend our clinics. By collecting drinking water samples from the students' home supplies (with parental permission), we have found an effective way to reach families who were not participating in our regular Extension program and to tie key messages to what students are learning in school.

Extension in Action: Engaging and Involving the Jamaica Bay Community after Hurricane Sandy

The area surrounding Jamaica Bay, New York City (NYC), forms a diverse mosaic of communities and households that span three counties, more than twenty neighborhoods, and twenty-three different local, state, and federal government jurisdictions (Sanderson et al., 2016). As communities struggle to recover from Hurricane Sandy, state and federal organizations and academic institutions have identified an added challenge to the recovery process: sustaining a healthy exchange of knowledge and solutions between residents, scientists, and agency personnel. New York Sea Grant (NYSG), in cooperation with the Science and Resilience Institute at Jamaica Bay (SRIJB), has developed an urban Cooperative Extension program in NYC, specifically for Jamaica Bay, to integrate knowledge and promote coastal resilience.[1]

Prior to Hurricane Sandy, few residents had experienced or knew the risk posed by large-scale climatic events on Jamaica Bay (Ramasubramanian et al., 2016). Hurricane Sandy was a wake-up call for New York City, indicating the city's vulnerability to extreme weather events. By hosting Climate Forums in the communities, SRIJB and NYSG aimed to enhance awareness of climate-related coastal events and provide tools to empower vulnerable communities and individuals. Holding the forums in the neighborhoods and working with community leaders have allowed NYSG and SRIJB to develop relationships with vulnerable coastal communities, such as Canarsie, a low- and middle-density community in southern Brooklyn that is generally characterized as a place for working-class families, where about one in nine people, ages sixteen and older, are unemployed. Some deemed Canarsie as an overlooked neighborhood when it came to Hurricane Sandy; for example, Canarsie was not mapped to be in the floodplain prior to the hurricane (NYC DCP, 2019). Yet the major storm flooded streets and destroyed homes.

After meeting with an active community member from Canarsie at a previous event, a block association president presented himself as a community champion, providing background and history about his community, giving a tour of the areas that were most heavily impacted by

Hurricane Sandy, overall demonstrating passion and defining needs for his community. This community member was excited and willing to work with SRIJB and NYSG to conduct a successful Climate Forum together in Canarsie to bring information, tools, and resources about how to handle and prepare for extreme weather events and climate adaptation. This partnership then led to a subsequent "Canarsie Listening Session." With funding provided to partners, Public Agenda,[2] SRIJB, and NYSG created a space for Canarsie residents to share their stories and experiences living in Canarsie and to hear from community members, civic and neighborhood leaders, and representatives of local public offices about how they are thinking about environmental and economic resiliency. Emergent themes included, but were not limited to, the importance of citizen networks for sharing information that residents can use and act on; the potential of youth leaders as catalysts for public work; and the need for meetings, events, and organizations that allow citizens to engage on a range of issues.

Throughout the process of building relationships in Canarsie and other Jamaica Bay communities, there have been challenges, including concerns about the ability and commitment to relaying information in relatable and culturally appropriate ways. To address this issue, including community leaders in both program development and future funding opportunities has helped provide a reputation with the communities as a trusted source. There is still a long way to go in terms of breaking down siloes between researchers, agencies, and communities, but the lessons learned to sustain engagement include

- building trust and giving value to community concerns and ideas,
- contributing to a collective planning process, and
- creating networks that enable people to better advocate for themselves and be more prepared for natural disasters.

The coproduction of knowledge by residents, researchers, and public agencies leads to more impactful application of science in communities. When the public is involved in a sustained way, people can better adapt and respond to changing ecological, physical, and social conditions.

Extension in Action: Collaborating with Southeast Alaskan Tribes to Sustain Culturally Important Practices and Resources in the Face of Climate Change

Climate change is a complex global phenomenon offering an uncertain future that is difficult to communicate to local audiences. In Alaska, climate change impacts vary across the state's diverse geography, from loss of Arctic sea ice, permafrost melt, and lack of winter snow in western Alaska to flooding, storm surges, ocean acidification, warmer waters, snowfall variations, invasive species, and toxins in the marine environment of southeast Alaska, all of which affect food security and access, and abundance of resources that are culturally important to Alaskan tribes.

Residents of rural communities in Alaska, especially along the coast, are experiencing changes in their environment that are noticeable in a single lifetime. In some parts of southeast Alaska, residents are observing the disappearance of yellow cedar, an important resource for cultural elaboration for Tlingit and Haida peoples (Buma et al., 2017; Hennon et al., 2016). Cedar is used in bentwood boxes, house posts, totem and mortuary poles, canoes, and masks as well as other cultural objects important for Tlingit and Haida identity. The close connection that indigenous people have with the environment for food, mobility, and ultimately survival makes them well positioned to observe climate change (Cochran et al., 2013; Krupnik & Jolly, 2002).

In southeast Alaska, tribal organizations led by the Central Council of the Tlingit and Haida Indian Tribes (CCTH) are leading their own initiatives to adapt to climate change. These efforts focus on identifying the climate stressors and potential impacts to key cultural resources, including salmon, shellfish, berries, yellow cedar, cultural sites, and human health. Effectively addressing these concerns requires partnering with federal and state agencies, the University of Alaska, nonprofits, and other local organizations to create strategies for monitoring change, develop mitigation strategies, and ultimately draft adaptation plans.

In 2016, Alaska Sea Grant facilitated a workshop that brought together tribes, researchers, agencies, and nongovernmental organizations to discuss impacts to these key cultural species and identify a path forward. During the workshop, local environmental coordinators from area tribes identified key monitoring efforts and partners to create or expand existing monitoring projects that could improve assessment and understanding of those culturally significant resources. These community-based monitoring efforts require technical experts from the University of Alaska, Sitka Tribe of Alaska, and local nongovernmental organizations, such as the Nature Conservancy and Southeast Alaska Watershed Coalition; land managers such as the U.S. Forest Service and the U.S. Fish and Wildlife Service, including government funders who also act as coordinators of public and private partnerships; and finally local experts in each of the tribes and nonprofits working in the region.

An example resulting from this effort includes a collaboration between Southeast Alaskan Tribes, the University of Alaska Fairbanks, U.S. Forest Service, Southeast Alaska Watershed Coalition, and Alaska Sea Grant. Several tribes in southeast Alaska were already monitoring rivers and streams important for salmon escapement by documenting temperature, dissolved oxygen, and other factors. To improve their ability to assess this culturally important food source, scientists created a model that allows tribes to upload monitoring data and receive potential futures for salmon populations in specific streams in southeast Alaska. This project was funded by Alaska Sea Grant, which saw the importance of how this collaborative project was developed. In 2019, the model was reviewed by tribes at a follow-up workshop facilitated by Alaska Sea Grant. Tribes also reviewed a regional adaptation plan developed by the CCTH at this workshop. Alaska Sea Grant has since provided climate data and decision support tools through individualized community workshops in southeast Alaska so that tribes can apply the regional adaptation plan to the critical needs they identify for their own communities.

Climate change has and will continue to impact coastal community residents in Alaska. Adapting to these changes can provide an opportunity to empower communities, build capacity, and improve community well-being and sustainability. Extension services, like those delivered by Alaska Sea Grant, provide a means of dialogue and data sharing, while retaining the power and decision-making in the hands of tribes. Alaska Natives have been adapting to climate change and changing sea levels for thousands of years. Knowledge and observations made by rural residents of Alaska of the changing climate can contribute to adaptation activities and provide feedback for scientific observations creating a dialogue with climate scientists that will ultimately lead to the adaptation solutions of the future.

Conclusion

Examples of environmental injustice are widespread, and the implications to the health and quality of life of individuals and communities affected are significant. Acknowledgement of this issue led to a 1994 executive order for all federal agencies to implement environmental justice policies and programs for minority and low-income populations. The U.S. Department of Agriculture (USDA) has sought to implement environmental justice practices in their mission with a variety of assistance programs focused on communities experiencing environmental injustice. AgrAbility, Outreach and Assistance for Socially Disadvantaged and Veteran Farmers and Ranchers Program, Federally-Recognized Tribes Extension Program (FRTEP), and support for 1890 and 1994 land-grant institutions are just a few examples of major programs focused on addressing environmental injustice. In their Environmental Justice Strategic Plan 2016–2020 (USDA, 2014), the USDA has committed to fully integrating EJ into technical and financial assistance programs and increasingly is committing resources to climate change issues such as flooding and drought.

Extension has the distinction and tradition of providing solution-based education to the masses, initially targeting the agriculture community and progressively reaching out to address expanding needs and emerging issues of an increasingly diverse population. Natural resources and environmental education are recognized as among the major core programs and successes within Extension. Educators are increasingly acknowledging the critical issues of environmental justice, yet a fundamental working knowledge of EJ and relevant education methods is lacking. Community-based participatory research, as described in this chapter, is a fundamental approach that could serve to maximize community engagement and beneficial societal impacts. Though some Extension educators use CBPR for various programming, it is not widely implemented, and therefore a universal challenge exists for Extension. How do Extension educators, primarily subject experts, become grounded in EJ literacy and educational approaches?

Opportunities to learn about, engage in, and build capacity for environmental justice are offered by a host of government agencies, scientific associations, and nonprofits across the nation. The federal government supports environmental justice and equity issues through conferences, workshops, webinars, partnerships, resources, and funding programs (U.S. Department of Energy, n.d.; EPA, 2019a, 2019b; National Institute of Environmental Health Sciences, 2021). The American Public Health Association offers both resources and webinars on healthy equity, racism and health, climate change, racial equity, and other topics relating to environmental health and justice (American Public Health Association, n.d.). And a host of nonprofits and research centers provide a variety of opportunities to learn from and network with environmental justice leaders and advocates at a national and regional level (Center for Community Engagement, Environmental Justice and Health, n.d.; Center for Earth, Energy and Democracy, n.d.; Center for Health, Environment, and Justice, n.d.; Citizen Science Association, n.d.; Deep South Center for Environmental Justice, n.d.; EPA, 2021; Low Country Alliance for Model Communities, n.d.;

National Black Environmental Justice Network, n.d.; North Carolina Environmental Justice Network, n.d.; WE ACT, n.d.).

Whether through partnerships, trainings, or consultation, any of these organizations can help Extension develop the knowledge, skills, and institutional capacity to address environmental justice, not only at the national level, but more importantly, at the community level through state programs. Fortunately, there are numerous existing support mechanisms within the national Extension framework for providing EJ and CBPR learning and professional development opportunities. Among these opportunities are the Joint Council of Extension Professionals (JCEP), a multitude of Extension program-related professional associations, and the National Association of Extension Program and Staff Development Professionals (NAEPSDP), which specifically provides support to Extension professionals who lead program and staff development initiatives. Further, in their summary of interviews with land-grant university presidents and other leaders, Gavazzi and Gee (2018) suggest that providing an undergraduate course in Extension not only would benefit students in understanding the land-grant system and its mission, but also enhance servant leadership and citizenship skills.

The opportunity and challenge exists for these Extension organizations to embrace EJ and CBPR as key goals in providing professional development opportunities for their members. Several suggested strategies for organizations to adopt EJ and CBPR include assessing their members' understanding of EJ and CBPR, and what information/skills they need related to these topics; committing to develop/organize training events, for example, webinars on EJ and CBPR for their members and Extension community based on the needs assessments; and incorporating EJ as a new priority in their strategic plan and mission. With increasingly tighter budgets, limited resources, and expanding needs, Extension needs to yet again rise to the occasion, work smarter, and consider environmental and social justice along with proven community education tools as a foundational skillset of educators to fulfill our mission.

NOTES

1. The Science and Resilience Institute at Jamaica Bay is a partnership between the City of New York, the National Park Service, and the City University of New York acting on behalf of a research consortium including seven other research universities and institutions, including New York Sea Grant. New York Sea Grant is a cooperative program of the State University of New York (SUNY), Cornell University, and the National Oceanic and Atmospheric Administration.
2. Public Agenda is a national, nonpartisan, nonprofit research and public engagement organization.

REFERENCES

American Public Health Association. (n.d.). Environmental Health. https://www.apha.org/Topics-and-Issues/Environmental-Health.

Anguelovski, I., Connolly, J. J. T., Pearsall, H., Shokry, G., Checker, M., Maantay, J., Gould, K., Lewis, T., Maroko, A., & Roberts, J. T. (2019). Opinion: Why Green "Climate Gentrification" Threatens Poor and Vulnerable Populations. *Proceedings of the National Academy of Sciences*, 116(52), 26139–26143.

Benham, B., Ling, E., Ziegler, P., & Krometis, L. (2016). What's in Your Water? Development and Evaluation of the Virginia Household Water Quality Program and Virginia Master Well Owner Network. *Journal of Human Sciences and Extension*, 4(1), 123–138.

Bodenreider, C., Wright, L., Barr, O., Xu, K., & Wilson, S. (2019). Assessment of Social, Economic, and Geographic Vulnerability Pre- and Post-Hurricane Harvey in Houston, Texas. *Environmental Justice*, 12(4), 182–193.

Bryant, B. (Ed.). (1995). *Environmental Justice: Issues, Policies, and Solutions*. Washington, DC: Island Press.

Bryant, B., & Mohai, P. (1992). *Race and the Incidence of Environmental Hazards: A Time for Discourse*. Boulder, CO: Westview Press.

Bullard, R. D. (1994). *Dumping in Dixie: Race, Class and Environmental Quality*, 2nd ed. Boulder, CO: Westview Press.

Bullard, R. D., Mohai, P., Saha, R., & Wright, B. (2007). *Toxic Wastes and Race at Twenty, 1987–2007: Grassroots Struggles to Dismantle Environmental Racism in the United States*. Cleveland, OH: United Church of Christ.

Buma, B., Hennon, P. E., Harrington, C. A., Popkin, J. R., Krapek, J., Lamb, M. S., Oakes, L. E., Saunders, S., and Zeglen, S. (2017). Emerging Climate-Driven Disturbance Processes: Widespread Mortality Associated with Snow-to-Rain Transitions across 10 Degrees of Latitude and Half the Range of a Climate-Threatened Conifer. *Global Change Biology*, 23(7), 2903–2914.

CCPH (Community-Campus Partnerships for Health). (n.d.). *Developing and Sustaining Community-Based Participatory Research Partnerships: A Skill-Building Curriculum*. https://www.cbprcurriculum.info/.

Center for Community Engagement, Environmental Justice, and Health. (n.d.). The University of Maryland Symposium on Environmental Justice and Health Disparities. https:// https://www.ceejh.center/umd-ej-symposium.

Center for Earth, Energy and Democracy. (n.d.). Tools. http://ceed.org/tools/.

Center for Health, Environment, and Justice. (n.d.). Leadership Training Academy. http://chej.org/take-action/help/leadership-training-academy/.

Citizen Science Association. (n.d.). Environmental Justice Working Group. https://citizenscience.org/get-involved/working-groups/ej-working-group/.

Clinical and Translational Science Awards Consortium. 2011. CTSA Community Engagement Key Function Committee Task Force on the *Principles of Community Engagement* (Second Edition). In *Principles of Community Engagement*, 2nd ed., NIH Publication No. 11-7782 (iv–v). Washington, DC: NIH. www.atsdr.cdc.gov/communityengagement/pdf/PCE_Report_508_FINAL.pdf.

Cochran, P., Huntington, O. H., Pungowiyi, C., Tom, S., Chapin, F. S., Huntington, H. P., Maynard, N. G., & Trainor, S. F. (2013). Indigenous Frameworks for Observing and Responding to Climate Change in Alaska. *Climatic Change*, 120(3), 557–567.

Corburn, J. (2005). *Street Science: Community Knowledge and Environmental Health Justice*. Cambridge, MA: MIT Press.

Craun, G. F., Brunkard, J. M., Yoder, J. S., Roberts, V. A., Carpenter, J., Wade, T., Calderon, R. L., Roberts, J. M., Beach, M. J., & Roy, S. L. (2010). Causes of Outbreaks Associated with Drinking Water in the United States from 1971 to 2006. *Clinical Microbiology Review*, 23(3), 507–528. https://cmr.asm.org/content/23/3/507.

Deep South Center for Environmental Justice. (n.d.). Our Work. https://www.dscej.org/our-work.

EPA (U.S. Environmental Protection Agency). (2016). Climate Change, Health, and Environmental Justice. EPA 430-F-16-054. https://www.cmu.edu/steinbrenner/EPA%20Factsheets/ej-health-climate-change.pdf.

———. (2019a). Centers of Excellence on Environmental Health Disparities Research. July 19. https://www.epa.gov/research-grants/centers-excellence-environmental-health-disparities-research.

———. (2019b). Environmental Justice Learning Center. August 13. https://www.epa.gov/environmentaljustice/environmental-justice-learning-center.

———. (2021). Environmental Justice in Your Community. March 9. https://www.epa.gov/environmentaljustice/environmental-justice-your-community.

EPA (U.S. Environmental Protection Agency), Office of Environmental Justice (OEJ). (2013). Environmental Justice-Related Terms as Defined across the PSC Agencies. https://www.epa.gov/sites/production/files/2015-02/documents/team-ej-lexicon.pdf.

Gavazzi, S. M., & Gee, E. G. (2018). *Land-Grant Universities for the Future: Higher Education for the Public Good.* Baltimore: Johns Hopkins University Press.

Gibson, J. M., & Pieper, K. (2017). Strategies to Improve Private-Well Water Quality: A North Carolina Perspective. *Environmental Health Perspectives*, 125(7), 076001-1-9.

Heaney, C. D., Wilson, S. M., & Wilson, O. R. (2007). The West End Revitalization Association's Community-Owned and -Managed Research Model: Development, Implementation, and Action. *Progress in Community Health Partnerships: Research, Education, and Action*, 1(4), 339–349.

Heaney, C., Wilson, S., Wilson, O., Cooper, J., Bumpass, N., & Snipes, M. (2011). Use of Community-Owned and -Managed Research to Assess the Vulnerability of Water and Sewer Services in Marginalized and Underserved Environmental Justice Communities. *Journal of Environmental Health*, 74(1), 8–17.

Hennon, P. E., McKenzie, C. M., D'Amore, D. V., Wittwer, D. T., Mulvey, R. L., Lamb, M. S., Biles, F. E., & Cronn, R. C. (2016). A Climate Adaptation Strategy for Conservation and Management of Yellow-Cedar in Alaska. U.S. Forest Service, Pacific Northwest Research Station, General Technical Report PNW-GTR-917.

Israel, B. A., Eng, E., Schulz, A. J., & Parker, E. A. (Eds.). (2005a). *Methods in Community-Based Participatory Research.* San Francisco: Jossey-Bass.

Israel, B. A., Parker, E. A., Rowe, Z., Salvatore, A., Minkler, M., Lopez, J., Butz, A., Mosley, A., Coates, L., Lambert, G., Potito, P. A., Brenner, B., Rivera, M., Romero, H., Thompson, B., Coronado, G., & Halstead, S. (2005b). Community-Based Participatory Research: Lessons Learned from the Centers for Children's Environmental Health and Disease Prevention Research." *Environmental Health Perspectives*, 113(10), 1463–1471.

Israel, B. A., Schultz, A. J., Parker, E. A., & Becker, A. B. (1998). Review of Community-Based Research: Assessing Partnership Approaches to Improve Public Health. *Annual Review of Public Health*, 19, 173–202.

Katner, A., Pieper, K. J., Lambrinidou, Y., Brown, K., Hu, C.-Y., Mielke, H. W., & Edwards, M. A. (2016). Weaknesses in Federal Drinking Water Regulations and Public Health Policies that Impede Lead Poisoning Prevention and Environmental Justice. *Environmental Justice*, 9(4), 109–117.

Keenan, J. M., Hill, T., & Gumber, A. (2018). Climate Gentrification: From Theory to Empiricism in Miami-Dade County, Florida. *Environmental Research Letters*, 13(5), 054001.

Krupnik, I., & Jolly, D. (2002). *The Earth Is Faster Now: Indigenous Observations of Arctic Environmental Change*. Fairbanks: Arctic Research Consortium of the United States.

Leker, H. G., & Gibson, J. M. (2018). Relationship between Race and Community Water and Sewer Service in North Carolina, USA. *PloS one*, 13(3). https://doi.org/10.1371/journal.pone.0193225.

Low Country Alliance for Model Communities. (n.d.). Environmental Justice. https://lamcnc.org/programs-initiatives/environmental-justice/.

Maupin, M. A., Kenny, J. F., Hutson, S. S., Lovelace, J. K., Barber, N. L., & Linsey, K. S. (2014). Estimated Use of Water in the United States in 2010. U.S. Geological Survey Circular 1405. 56 p.

Minkler, M., & Wallerstein, N. (2003). *Community-Based Participatory Research for Health*. San Francisco: Jossey-Bass.

National Black Environmental Justice Network. (n.d.). Our Work. https://www.nbejn.com/ourwork.

National Institute of Environmental Health Sciences. (2021). Partnerships for Environmental Public Health. April 26. https://www.niehs.nih.gov/research/supported/translational/peph/index.cfm.

North Carolina Environmental Justice Network. (n.d.). Get Involved with NCEJN Events. https://ncejn.org/events/.

Northridge, M. E., Yankura, J., Kinney, P. L., Santella, R. M., Shepard, P., Riojas, Y., Aggarwal, M., & Strickland, P. (1999). Diesel Exhaust Exposure among Adolescents in Harlem: A Community-Driven Study. *American Journal of Public Health*, 89(7), 998–1002.

NYC DCP (New York City Department of City Planning). (2019). Resilient Neighborhoods: Canarsie. https://www1.nyc.gov/assets/planning/download/pdf/plans-studies/resilient-neighborhoods/canarsie/summary-report-canarsie.pdf?r=1.

Pieper, K. J., Krometis, L. A., Benham, B. L., Gallagher, D. L., & Edwards, M. (2015). Incidence of Waterborne Lead in Private Drinking Water Systems in Virginia. *Journal of Water and Health*, 13(3), 897–908.

Ramasubramanian, L., Menser, M., Rieser, E., Brezin, M., Feder, L., Forrester, R., Allred, S., Ferenz, G., Bolstad, J., Meyer, W., & Tidball, K. (2016). Neighborhood and Community Perspectives of Resilience in the Jamaica Bay Watershed. In E. W. Sanderson, W. D. Solecki., J. R. Waldman, & A. S. Parris (Eds.), *Prospects for Resilience: Insights from New York City's Jamaica Bay* (pp. 117–137). Washington, DC: Island Press.

Sanderson, E. W., Solecki, W. D., Waldman, J. R., & Parris, A. S. (Eds.). (2016). *Prospects for Resilience: Insights from New York City's Jamaica Bay*. Washington, DC: Island Press.

Schaider, L. A., Swetschinski, L., Campbell, C., & Rudel, R. A. 2019. Environmental Justice and Drinking Water Quality: Are There Socioeconomic Disparities in Nitrate Levels in U.S. Drinking Water? *Environmental Health*, 18(3). https://doi.org/10.1186/s12940-018-0442-6.

Stillo, F., & Gibson, J. M. (2018). Racial Disparities in Access to Municipal Water Supplies in the American South: Impacts on Children's Health. *International Public Health Journal*, 10(3), 309–323.

Switzer, D., & Teodoro, M. (2017). The Color of Drinking Water: Class, Race, Ethnicity and Safe Drinking Water Act Compliance. *Journal of American Water Works Association*, 109(9), 40–45.

UCC (United Church of Christ), Commission for Racial Justice. (1987). *Toxic Wastes and Race in the United States: A National Report on the Racial and Socioeconomic Characteristics of Communities with Hazardous Waste Sites*. New York: Commission for Racial Justice.

USDA (U.S. Department of Agriculture). (2014). Environmental Justice Strategic Plan 2016–2020. 28 pp. https://www.dm.usda.gov/emd/responserestoration/docs/8162572_USDA%20EJ%20Strategy%20 Final.pdf.

U.S. Department of Energy. (n.d.). Environmental Justice: Office of Legacy Management. https://www.energy.gov/lm/services/environmental-justice.

Virginia Cooperative Extension. (2019). Virginia Household Water Quality Program. April 21. https://www.wellwater.bse.vt.edu.

WE ACT. (n.d.). Environmental Health and Justice Leadership Training. https://www.weact.org/ home-3-2/getinvolved/education/ehjlt/.

Wilson, S. M. (2009). An Ecologic Framework to Study and Address Environmental Justice and Community Health Issues. *Environmental Justice*, 2(1), 15–24.

———. (2010). Environmental Justice Movement: A Review of History, Research, and Public Health Issues. *Journal of Public Management & Social Policy*, 16(1), 19–50.

Wilson, S. M., Heaney, C. D., Cooper, J., & Wilson, O. (2008a). Built Environment Issues in Unserved and Underserved African-American Neighborhoods in North Carolina. *Environmental Justice*, 1(2), 63–72.

Wilson, S. M., Richard, R., Joseph, L., & Williams, E. (2010). Climate Change, Environmental Justice, and Vulnerability: An Exploratory Spatial Analysis. *Environmental Justice*, 3(1), 13–19.

Wilson, S. M., Wilson, O. R., Heaney, C. D., & Cooper, J. (2008b). Community-Driven Environmental Protection: Reducing the P.A.I.N. of the Built Environment in Low-Income African-American Communities in North Carolina. *Social Justice in Context*, 3, 41–58.

Wing, S. (2002). Social Responsibility and Research Ethics in Community-Driven Studies of Industrialized Hog Production. *Environmental Health Perspectives*, 110(5), 437–444.

Wing, S., Cole, D., & Grant, G. 2000. Environmental Injustice in North Carolina's Hog Industry. *Environmental Health Perspectives*, 108(3), 225–231.

Programming to Address Building Financial Equity

Katherine E. Soule and Derrick Robinson

Educators in financial literacy often teach clientele core concepts to help them save money, develop budgets, and make investments. These core concepts generally fall within the following categories: financial planning (including budgeting and savings), debt management, understanding credit agreements, and how-to investment. This type of education focuses on individuals' financial knowledge and habits in order to help individuals manage their finances responsibly. Within the financial education framework, *financial equity* is defined as "the difference between the value of an individual's assets and liabilities. It is also referred to as 'net worth.' Financial equity is more commonly called 'equity' . . . the term financial equity [is used] to distinguish it from the use of equity as it relates to fairness" (Higher Rock Education, 2018).[1] Financial educators work to increase individual consumers' financial literacy, promoting positive fiscal decisions and habits.

This chapter takes a broader view, recognizing that individuals' financial situations are embedded within the larger social structures of our society. These social structures are made up of policies, systems, environments, and social norms that impact the financial situations of individuals.

When educators and service providers consider only individuals' circumstances, they are likely to overlook how social and environmental factors impact, and in some cases predetermine, individuals' financial decisions (cf. Shildrick & Rucell, 2015). In this chapter, we use the term *financial equity* to mean that everyone has a fair and just opportunity to access resources (i.e., land, financial capital, labor, etc.) in order to increase financial and economic capability and security. To do so requires removing institutional and societal obstacles such as community-level poverty, discrimination in financial services access, fair wages across comparable occupations, quality public education, affordable housing, and safe environments. Each of these obstacles has some influence over individuals' economic and financial well-being. Specifically, these influences have been regulated as a function meant to benefit certain groups (i.e., individuals or organizations that already have capital resources and/or accumulated wealth and have a self-interest in maintaining the structures that have facilitated the growth of their resources and/or wealth) over others (i.e., individuals, communities, and organizations that have historically faced obstacles and barriers to economic markets). Understanding how these various levels of influence impact individuals' financial literacy and decisions can help educators develop and deliver more effective programs centered in increasing economic well-being of our most marginalized people and communities.

Overview of Sociology and Personal Economics

Economists tend to view consumers' experiences through "free choice models," where rational consumers make their own financial decisions and are thereby responsible for their financial successes and struggles (Piachaud, 1987). Consequently, much of financial literacy education focuses on increasing consumers' financial knowledge with the goal of increasing consumers' financial decision-making capacity toward improving their fiscal circumstances. Despite educators' efforts to support individuals' financial decision making and behaviors, researchers evaluating such financial literacy education programs have shown few positive effects of program participation (Amagir et al., 2017; Lusardi, Michaud & Mitchell, 2017; Gale & Levine, 2010). By exploring individuals' fiscal decisions and circumstances within broader contexts, new understandings emerge, which could assist financial educators and practitioners. In fact, the Canadian government has determined a "holistic approach" that considers contextual and environmental situations is the "most effective" approach to financial education (Financial Consumer Agency of Canada, 2018). Through a sociological lens, researchers explore poverty in relationship to culture, power, and social structures. Frequently, sociologists also consider how the roles of race/ethnicity, gender, and other socio/cultural characteristics can impact individuals' circumstances. Inherent in these considerations is the intersectionality of wide-ranging life circumstances: hunger, access to health care, educational environments, violence, family structure, and so on. Sociological perspectives can highlight how financial decisions and behaviors are entwined with one's life context, not simply a measure of one's capacity to be fiscally responsible.

Historical Impacts on Financial Equity

Historical biases that led to discriminatory practices and institutions have profound impacts on individuals' financial situations today. Birkenmaier and Sherraden provide several examples of these types of practices:

> The forced loss of land and removal to reservations continue to influence the financial positions of Native Americans. Current wealth holding among African Americans cannot be understood without appreciating property relations under slavery and the historical exclusion of African Americans from property ownership. In the 1930s, Mexican Americans—many of them U.S. citizens—were forcibly repatriated to Mexico. The loss of assets has affected the wealth of subsequent generations of Hispanics. In contrast, White families received free land through the Homestead Act of 1862, which has been the largest asset distribution policy in U.S. history. Homestead lands laid a foundation of wealth for contemporary White families, an economic advantage denied to minority families. (Birkenmaier & Sherraden, 2018, para. 5)

Such historical practices created systems of privilege for white families in America that continue to impact their descendants' access to wealth and fiscal resources through inheritance, family legacies, and generational knowledge of financial institutions. Likewise, communities who suffered from these practices continue to experience wealth inequity and limited access to financial systems today.

Women have also experienced negative fiscal impacts from historical practices that prevented and/or limited their fiscal agency. For example, women were not granted rights to retain their own property until 1839. Women could legally be paid less than men for the same work until 1963. According to the Institute for Women's Policy Research (2021, para. 1), in 2018 women working full-time made 82 cents on the dollar compared to men who were employed full-time. Findings from a 2018 MassMutual study indicated that women were less likely to save for retirement, pay down debt, or consider themselves to be the primary decision maker for family financial planning than men (Halloran, 2018). Women's current financial situations and habits reflect historical practices that limited women's roles in making money and managing their own finances. The historical inequalities that women and marginalized cultural groups have experienced in the United States not only continue to impact individuals' fiscal resources, they can also impact one's financial self-efficacy.[2]

The Role of Self-Efficacy in Financial Equity

Self-efficacy is defined as individuals' belief in their own capacity to achieve goals (Bandura, 1977). To succeed in implementing financial knowledge and positive fiscal behaviors, a consumer "needs a sense of self-assuredness, or 'self-belief,' in their own capacities" (Farrell, Fry & Risse, 2016, p. 85). In an eight-year longitudinal study of young adults, researchers found that millennial men had higher financial self-efficacy than women (Serido & Shim, 2017). Many people in marginalized populations have low self-efficacy as a result of being excluded from fully accessing opportunities and resources available to those in privileged social positions (Minnesota Psychological Association, 2018). In his article on the role of the neighborhood as cultural context for self-efficacy and academic achievement, Merolla describes a common experience in impoverished neighborhoods:

> Many individuals within impoverished neighborhoods lack access to stable employment and decent wages, which erodes their confidence that their actions can positively influence their socioeconomic situation. For example, Wilson cites individuals who hold long-term employment in low-wage jobs, yet remain in economically precarious situations—often only a single adverse event (e.g., serious illness) away from falling into abject poverty. For these individuals, their sense of powerlessness is a product of a lack of access to the resources and social institutions that enable individuals to overcome difficult circumstances.
>
> However, at the same time, individuals in disadvantaged neighborhoods also understand and acknowledge dominant cultural ideologies that link hard work and individual success. For instance, Wilson (1996) notes that despite facing structural constraints to economic mobility, many residents in poor neighborhoods verbally reinforce basic American values in terms of the importance of hard work and individual initiative in trying to get ahead economically. (Merolla, 2017, p. 380)

This experience occurs throughout the country, in urban and rural communities. Awareness of the impacts of social context on self-efficacy points to gaps in traditional financial education, which strives to teach consumers positive fiscal behaviors and decision making. Even with knowledge of these positive practices, consumers with low self-efficacy experience greater challenges in implementing the knowledge. Merolla's description highlights the need to address self-efficacy, not only factors that create these social contexts. His description also hints at another issue: the need to consider the impacts of culture and acculturation in financial education.

Considering Cultural Influences

Many people's fiscal values and behaviors are rooted in cultural practices and beliefs about finances. For example, according to Islamic principle, Muslims are not allowed to benefit from loaning someone money or from receiving money from someone else. In other words, Muslims do not believe in paying interest. As a result, in the United States where paying interest is a common fiscal practice, Muslims would have greater difficulty in obtaining property, starting a business, and building a credit score than those whose belief systems allow them to pay interest. In Latino communities, there is a great value on shared living and social interdependence. These values mean that family members will often use money to support one another's needs and goals, as opposed to saving funds for later personal use, such as retirement. Such cultural differences toward fiscal practices impact individuals' decisions about how to spend and save money. Most financial education in the United States reflects the values and practices of dominant American culture, which can further marginalize those cultural groups who are often not reflected in mainstream contexts.

At the same time, as pointed out in Merolla's description earlier, "many residents in poor neighborhoods verbally reinforce basic American values in terms of the importance of hard work and individual initiative in trying to get ahead economically." While Merolla did not address the race/ethnicity, religion, or gender of those in the neighborhood he described, it is common for individuals of marginalized cultures to experience some level of acculturation. Collier, Brice, and Oades-Sese defined acculturation as

> a type of culture change that occurs when an enculturated individual comes into the proximity of one or more new or different cultures or subcultures. This may occur by moving into a new environment or location populated by people raised with a different language or culture. It may occur by going to a new school or moving to a new region of the country. It may also occur through exposure to movies, television, and books. (Collier, Brice & Oades-Sese, 2007, p. 356)

As individuals experience acculturation, they may begin to experience changing values and beliefs. Yet Collier, Brice, and Oades-Sese emphasized that "most cultural groups do not give up valued practices lightly, whether economic, religious, or communicative" (2007, p. 355). Individuals may experience conflict between new and culturally ingrained beliefs about money habits and financial values.

An understanding of social contexts, such as historical or cultural influences, can be useful for those considering the effectiveness and relevance of programs to address financial equity.

In the next section, we will consider several financial literacy programs and services through a social (and economic) justice perspective.

Is Current Financial Literacy Programming Addressing Financial Equity?

Household management and financial literacy education programs have integrated content across topic areas and become the primary means for providing education meant to increase financial security. These programs are grounded in the theory that increased financial education leads to greater financial knowledge, which then leads to better financial behavior, and ultimately leads to improved consumer outcomes and well-being (Hathaway & Khatiwada, 2008). Therefore, development of effective program evaluation tools is essential toward capturing any changes in participant financial literacy (Lyons et al., 2006). Among evaluated programs, there are both effective (Boyce & Danes, 1998; Danes, 2005; Varcoe et al., 2005; Loke, Choi & Libby, 2015; Kim, Anderson & Seay, 2018) and less effective (Gartner & Todd, 2005; Jump$tart Coalition for Personal Financial Literacy, 2006; Mandell, 2008; Mandell & Klein, 2009) results. One issue with these programs is the evaluation tools, and whether they are effectively measuring changes in financial literacy and knowledge accumulation or changes in personal financial behaviors without retention of purpose behind the behavior change (Hastings, Madrian & Skimmyhorn, 2013). Notwithstanding these program developments, Lusardi and Mitchell (2014) found most Americans are still financially illiterate. Therefore, programs offered have to consider issues around sociodemographics and other influential socioenvironmental characteristics that affect financial literacy.

Since a large portion of the population is not literate about basic financial concepts, it is highly improbable that short-term financial literacy training would significantly impact consumers' decision-making capacity over time (Gale, Harris, & Levine, 2012). Additionally, Fernandes, Lynch, and Netemeyer (2014) found the effectiveness of financial literacy education on financial behaviors to be extremely small and diminishing over time, especially for programs targeting low-income populations. Low-income populations are often the most marginalized as it relates to financial equity. These realities force us to consider how much current financial literacy programming perpetuates the structural and institutional barriers that impact financial inequity.

Current programs seem to engage deficits in financial knowledge by providing resources focused on improving financial skills, meaning the programmatic outcomes become centered around improving areas of clientele capability in areas identified as necessary to increase financial security (i.e., developing savings strategies, reviewing transactions, household budgeting, etc.). However, many of these programs do not account for employment-related inequality that is correlated with low wages. For instance, low-wage jobs are associated with lack of access to traditional savings channels offered through individuals' employers (i.e., 401k, health savings accounts, etc.). Lack of access to these channels makes it difficult for heuristic behavioral nudges to be effective, leaving individual participants to incorporate learned savings strategies as the primary means of accumulating assets. Overlooking these types of shortcomings helps to perpetuate the status quo: providing cheap labor for capitalist-based business ventures without accounting for social and community costs associated with inadequate asset accumulation across ages.

According to Gale and Levine (2010), another shortcoming of financial literacy and household management education programs is a decrease in the number of formally trained family and consumer sciences educators. These educators are often the primary facilitators of providing public human capital investment into education programming with the intent to increase financial equity. Werhan (2013) found the number of teachers available for financial literacy education in high schools and middle schools has decreased 26 percent since the early 2000s. Student enrollment in financial literacy education also decreased 38 percent over the same time period.

Perhaps this situation actually presents an opportunity for Cooperative Extension professionals to change the methodology of financial literacy education. More specifically, Cooperative Extension professionals may find that this is the time to develop program resources that integrate financial literacy education across disciplines covered in current high school and middle school curricula, especially when engaging traditionally marginalized communities with different cultural norms (Cordero & Pedraja, 2019). This change in financial literacy education could help disrupt structural barriers and, ultimately, increase financial literacy lessons for students from low-income and other marginalized communities.

Beyond working in partnership with public and private schools, community schools, and community colleges, Extension professionals have unmet opportunities to engage with other youth service providers such as social workers, community educators, and faith-based organizations to support increased opportunities for youth to participate in financial literacy education. For example, although social workers are considered on the frontlines of providing and advocating for economic justice opportunities for underserved and marginalized populations, many do not have the financial knowledge and skills to address household finances and financial practice (Sherraden, Frey & Birkenmaier, 2016).[3] They traditionally gain these skills directly on the job or through continuing education and professional development trainings during employment periods. Providing financial literacy education and training to social workers and other community-based youth service providers will enhance community capacity to support increases in youth financial literacy.

Developing Skills for Life Management on a Minimum Wage

The push for equitable pay has been an ongoing struggle in the United States. In 1968, Martin Luther King Jr. recognized fair wages to be a civil right afforded to all citizens. Yet still, fifty years after his assassination, the fight for equitable wages for the lowest paid workers is still ongoing. While evidence has shown positive socioeconomic benefits from increasing minimum wage earnings, this does not address the structural issues that make minimum wage laws necessary in the first place (Van Dyke et al., 2018; Cooper, 2019). Notwithstanding those structural issues, family and consumer science Extension professionals have made gains in providing sustenance-based management techniques. This has led to programmatic efforts being developed that focus on teaching household resource management survival strategies. However, these programs do little to provide knowledge on asset accumulation and other techniques focused on building net worth. This is potentially an issue that needs to be addressed with the training of those responsible for facilitating financial education programs. Loke, Birkenmaier, and Hageman (2017) found that facilitators' limitations in coverage of finances and asset building was mainly a result of the facilitators' perception of their own asset-building capacity and practice. If this is correct, then facilitator effectiveness to increase financial literacy and financial decision-making outcomes of their clientele lies in the facilitators' self-efficacy about being financially literate. Therefore, only those explicitly trained and practicing effective asset building and financial decision making (i.e., professional financial investor at a bank) are capable of providing education effectively across financial literacy curriculums.

However, merely subsisting does not address the real issue: the need for a pathway out of poverty for individuals, households, or communities. The purpose for any publicly funded financial literacy programming is to increase participants' socioeconomic well-being, especially for low-income households. The current gaps in programmatic opportunities create incentives for program providers to focus on developing new programs that promote participants' capacity toward increased financial and economic equity across sociodemographic characteristics and diverse consumer preferences for spending and savings. According to Lusardi and Mitchell (2014), there are significant differences across financial education programs, and therefore in order to be effective, programs need to be targeted to specific groups. By providing access to different financial literacy education programs focusing on identifying life-cycle stage, social and environmental factors, and public services available, program providers and professionals can support individuals' decision-making optimization toward wealth accumulation.

Fostering Skills in Locating and Maintaining Affordable Housing

Along with consumer product decision making is the decision for housing. The decision on housing type and location is important for two reasons. First, housing location decides what other public goods and resources are available, as well as what types of policies are expected to impact an individual or household. Low-income individuals and households who would be expected benefactors from financial literacy education programs are traditionally concentrated in communities with issues of poverty. Concentrations of poverty affect types of schools available, types of public infrastructure, and availability of localized employment. Secondly, a large amount of people are housing insecure, meaning they are trading off spending in other areas (i.e., food, health care, childcare, etc.) to cover their housing needs. According to the Centers for Disease Control, housing insecurity impacts approximately 28–40 percent of adults who own or rent homes, with Black and Hispanic populations being the most affected (Njai et al., 2017). Across ages, this problem impacts those between twenty-five and thirty-four the most. The University of California has been trying to address this issue through the development of a financial literacy education resource guide focused on increasing financial knowledge around the decision to live independently made by youth and young adults.

Other programs that help to decrease housing insecurity and encourage homeownership, as well as home renovation and repair programs, are offered through the U.S. Department of Agriculture (U.S. Department of Agriculture, 2019). However, these programs focus primarily on simply minimizing household costs through ownership and renting and are concentrated in rural communities. While this program provides an increased socioeconomic benefit by decreasing the amount of defaults among renters, helping to maintain current property values, and increasing residential growth in some communities, the program does little to address the structural issues that led to the programming need. For groups who do not qualify, the programs can create further community inequity by withholding access to some of the most marginalized populations, who also may have the greatest need for the resources.

At the state level, multiple Extension agencies are providing programmatic development and opportunities around finding affordable housing. For example, the University of Georgia has multiple programs meant to help increase community and individual wealth around housing (University of Georgia, 2019). One program focuses on the individual by providing a "Homebuyer Education Course" that is expected to help optimize the decision making for the individual when choosing to purchase a home. In theory, homebuyers will identify their needs, what resources are required, and what are the best strategies to obtain and maintain a home. Another program offered through University of Georgia Extension focuses primarily on community housing, which engages multiple community residents and stakeholders to support them in optimizing the community development decisions for housing. A program offered through Iowa State University focuses on increasing community wealth by encouraging effective development around communities' needs through assessments (Wallace, 2017).

While these types of programs help to establish equity for individuals and communities who historically may have been marginalized in housing market access, these programs do not provide information around other strategies to accumulate assets. Some strategies that could help individuals accumulate assets include leveraging capital investments, developing limited liability corporations to leverage personal capital accumulation, developing community-based land trusts, using cooperative development models to accumulate and distribute equity, or strategies for optimally diversifying or creating investment portfolios. These are areas where the private sector becomes the main supplier of services and information around these topics. This limits access to channels for wealth accumulation for communities without enough accumulated capital to incentivize private market support. With inequitable access to these strategies, communities are unable to increase their financial equity. Therefore, more focus needs to be on making sure program facilitators have diverse program offerings and are comfortable with disseminating information in support of wealth accumulation strategies, especially understanding the social and environmental factors that influence financial decision making of marginalized communities.

Opportunities to improve the socioeconomic well-being of marginalized communities should not be limited to minimizing household production costs (Becker, 1965). Increasingly households are finding benefits in obtaining properties for the purpose of renting them to generate household revenues. This shows that not only are taste preferences of the property purchaser important, but also those of the property occupant. However, little programmatic attention has been given to supporting households in this form of business development.

Leveraging Workforce Development Programming toward Wealth Building

Workforce development programs have been reactive, responding to the ever-changing needs of local business owners. According to Mincer and Polachek (1974), human capital (education) investments in skills that increase productivity will provide a positive benefit to the individual, and potentially the entire household. Notwithstanding this theory, programs concentrated in providing low-skill, low-intensity human capital development may not have strong enough wage returns for increased productivity (i.e., culinary arts training, janitorial services, etc.) At the same time, employment in these sectors traditionally has paid lower wages. Increased employment without increased wages often leads to increased financial inequities. Some programs have shown high returns on investment across multiple programming types to community socioeconomic well-being. This could be related to the consistent offering of low-skill labor knowledge to program participants. While that helps to meet a direct community need, it doesn't help to provide future, innovative labor supply skills. It also is potentially perpetuating the low-skill labor supply concentrated in already marginalized communities.

Workers' perceptions and motivations for labor-force participation are associated with the self-efficacy ideas discussed in the previous section. This also aligns with research that illustrates low labor-force participation at the lower levels of income distribution. Many workers in these types of jobs may not be able to see themselves as beneficial inputs to the production function of the businesses offering employment. Combining this idea with potential cost/benefits related to household production savings from individuals providing caregiving opportunities, etc., one can see the importance of providing programs that increase skills and positive perceptions of labor skill value of the individual.

Alternatively, a workforce development program focused on improving community financial equity must provide skills that promote innovation, critical thought, and increased problem-solving within participants. An equitable program must also allow for increased community access and input during the planning and development stages of the programs. Communities want higher-wage job opportunities, along with strategies to accumulate and manage household assets.

Recent workforce development programs have begun focusing on high-skill labor capital development. Specifically, institutions have begun using "Extension" models for localized workforce development; for example, the University of California, San Diego Extension offers Python coding courses (UCSD Extension, 2019). These types of programs are expected to decrease frictional employment by offering educational certificate programs designed for employed and unemployed individuals, as well as to help encourage labor-force participation through active training.

However, investments into these programs by participants are budgetarily constrained, meaning individuals with the greatest need (i.e., low-income, unemployed labor-force participants seeking training for job placement) oftentimes are priced out of the programs. Even with programs offered through private firms, the economic cost-benefit leads employers to reserve these training opportunities to those with the highest skill, or higher expected return on investment. According to Heinrich (2013), even when costs are reduced or subsidized, other barriers (i.e., transportation costs, tradeoffs of employable hours, childcare costs, etc.) may prove restrictive. In the context of these issues, Heinrich suggests increased investment by public entities, private corporations, and individuals. Moreover, these programs should be multifaceted, combining basic, occupational, on-the-job, and sector-based skill development to fully engage productivity. To increase financial equity, programs need to include financial literacy resources, training for wage negotiation strategies, as well as managing personal financial investment strategies.

Destabilizing "Savior Complex/Culture" in Our Work

The population providing programmatic outreach has a responsibility to consider the lens or perspective of the deliverer. This is important to ensure that the behaviors and considerations that led to financial inequities are not continued and, more specifically, that programmatic resource delivery doesn't create norms that perpetuate the financial inequities of marginalized groups (i.e., using traditional savings vehicles that have helped contribute to financial inequity over time). Cooperative Extension professionals should be asking, "How do we facilitate and cocreate more with community-based solutions and increased community input from those expected to be impacted most?"

In order to establish financial equity in a community, community development projects meant to increase the socioeconomic well-being of local residents should include those local residents in the planning and decision-making processes. Many times large-scale development organizations coupled with exo-community landowners are responsible for the development patterns in marginalized communities. Therefore, it's important to provide marginalized communities with the tools to comprehend the development plans introduced and enacted within their communities, similar to University of Georgia's Initiative for Community Housing program. This program encourages equitable competition by local groups in community development projects through promoting placemaking and providing comprehensive engagement strategies across participating community groups. These types of opportunities are important toward promoting financial equity because they mitigate the structural and institutional barriers that traditionally have kept marginalized participants from being involved in the process. Other communities have engaged in placemaking and financial equity building through community/family/multigenerational property ownership where cooperative ownership models are used as a catalyst to accumulate equity for individuals and communities.

Leveraging Cooperative Extension to Build Financial Equity in Communities

Currently, Extension programs have focused on providing household resource management as a means for increasing individual financial capability. While this type of programming helps stabilize household budgets, it often lacks in strategies for exponentially increasing asset accumulation. These skills are especially important for households dealing with generational income disparity. Let's explore how one Cooperative Extension financial literacy program, the Money Talks Program, could address building financial equity (UC ANR, 2019).

Many of our existing programs could provide improvements by seeking to be inclusive across our clientele's needs. While it is important to address financial security and well-being within communities and households, Cooperative Extension should be providing program resources to increase community and household equity outside of saving cash. Using a multitiered approach, similar to those meant to increase outreach for college pathways, Extension has to expose more individuals to equity building programs. The multitier method expects that for every general workshop provided, groups of those who participated will continue to stay engaged at higher levels (i.e., online training, low intensity class, etc.). Of those higher-level participants, a proportion will increase their level of participation to some form of certificate program training or other formalized educational opportunity (i.e., postsecondary/graduate studies in family and consumer science). Using this approach allows Extension professionals to induct change in community-based sociocultural norms, helping individuals establish a positive attitude toward participation within programs, as well as extending and disseminating information passed to them through Extension educators, with a possibility of community-based change instituted around the highest reachable levels of financial equity program offerings. While Extension professionals may have expertise in financial literacy, needs vary across populations and communities; therefore, engaging the community in the planning and implementation of these opportunities is critical to the success and effectiveness of the efforts.

The Money Talks Program attempts to address this by exposing youth and young adults to financial literacy education materials, most recently around decreasing housing security. Money Talks uses traditional methods of providing educational resources, such as guides that include some background information and activities. These resources are used by community educators to help disseminate the financial literacy information regardless of the educators' perceived self-capacity and practice of strategies. Money Talks has also been using supplementary resources, such as informational videos and interactive online resources. However, all these strategies assume that participants have, or will have, some level of experience with the topics (i.e., banking, credit cards, etc.). This is not always the case. Also, many of the partners using the materials are not always able to reach those outside of previously existing instructional areas (i.e., schools, churches, club organizations, other nonprofits, etc.). Therefore, while the multitier approach may be effective, Extension professionals are not always including an exhaustive representation of clientele in the lowest tier. Money Talks could be more intentional toward decreasing financial inequities by ensuring that our outreach methods and resource development take into consideration the socioeconomic and environmental issues that influence the capacity for program participants to institute financial literacy information into everyday, long-term practices.

How to Increase Equity of Community's Financial Services Access

Extension professionals have roles to play in helping communities to reduce barriers to financial equity. For example, financial access deserts are an issue in our current landscape.[4] These areas are characterized as spaces where people have reduced access to financial capital, especially through traditional channels of access (i.e., banks). Access to financial capital is linked to increases in community wealth and well-being. However, without knowing where these deserts are located, policy makers are unable to make appropriate and effective interventions to reduce the likelihood of these areas continuing. Extension professionals are available to provide this information to policy makers through mapping technology and also providing advisement on how to reduce the areas affected by these deserts. Policy makers can use this information to target policy meant to increase financial institutions in an area (i.e., incentives to increase brick-and-mortar financial services locations) as well as consider offering alternative financial institutions (i.e., public banking options). As Extension professionals begin or continue to advise and advocate to policy makers, they are helping to reduce the financial equity barriers experienced by their clientele.

This information could also be provided to coalition groups who are involved in economic justice advocacy or local land development, as well as financial service providers themselves who may be unaware of their inequitable practices. For example, 4-H projects could be partnered with financial service providers to facilitate lending for farm animal purchases and maintenance. This type of program follows the multitier approach and could help reduce financial inequity traditionally experienced by certain 4-H participants in the past, while also instilling financial literacy education around asset building.

Maintaining Cultural Relevance While Building Financial Equity

Currently, Cooperative Extension professionals are beginning to assure resources are provided to culturally sensitive communities (i.e., non-English speaking households). However, these resources are not always culturally relevant in content while being available in languages other than English. A possible way to address these shortcomings is to ensure consideration of culturally heterogeneous family/community structures; for example, when offering savings strategies and examples, taking into consideration cultural norms for savings. Potentially, our one-size-fits-all approach provides further inequity by only representing a monocultured experience when describing behavioral savings models for individuals. Another example is the social interdependence of Latino communities, which means that family members will use money to support one another's needs. Such cultural differences toward fiscal practices impact individuals' decisions about how to spend and save money. Most financial education in the United States reflects the values and practices of the dominant American culture, which can further marginalize those cultural groups who are often not reflected in mainstream contexts. Therefore, to help increase financial equity in these communities, we could provide programs that focus on cooperative models that allow for cultural norms to be integrated into savings promotion strategies.

When considering ways to facilitate financial literacy training while promoting financial equity, we should consider incorporating placemaking when considering the spaces financial literacy education is offered. These ideals are community-based and leverage communities' assets and motivations for positive outcomes to create quality spaces for community members. According to the Project for Public Spaces (2007), people have a connection with placemaking spaces, and these spaces can be catalysts for community revitalization. This is important when considering equitable financial literacy program offerings, and assuring the offerings are community-based and a safe place for communities and individuals to share knowledge, experiences, and peer-to-peer learning. This is especially true for financial literacy education, because the information shared in these spaces can reveal personal vulnerabilities and in some cases are considered to be invasive (Willis, 2011). Therefore, alternative tools, spaces, and approaches need to be considered to ensure financial equity access by low-income and other marginalized individuals.

Ensuring culturally competent practices will be beneficial toward providing effective financial literacy education and equity. According to Birkenmaier and Sherraden (2018), cultural competency practices are essential to providing financial literacy education and equity. They suggest that program providers should be aware of their own values and beliefs, and how those influence the strategies they provide to clientele. For example, recognizing one's willingness to trust governmental institutions to provide services has some influence on the willingness to inform clientele about those resources. However, the same clientele participating in the program may feel marginalized by those same institutions that the program provider is attempting to build participants' trust for. They also suggest that program providers comprehend the effects institutions have had on shaping the current financial equity issues that are being addressed. Without understanding the holistic impact of the institutions, the program provider could relate participants to institutions that created their inequities, thereby making it difficult to institute any changes that could increase the financial well-being of program participants. Even more so, program providers need to understand how these institutional realities influence current access to financial products and services. Without incorporating these ideals into the practices that inform the delivered content across programs, these programs could essentially perpetuate the inequities they were intended to diminish. Therefore, financial literacy education programs need to be diverse across spaces of access, need to be community-based and consider culturally relevant practices that directly meet community needs, and need to be multifaceted, targeting distinct groups with specific products relevant to their needs and capacity.

NOTES

1. Financial Equity: Everyone has a fair and just opportunity to access resources (i.e., land, financial capital, labor, etc.) in order to increase financial and economic capability and security.
2. Financial Self-Efficacy: An individual's sense of their capacity to successfully manage their finances and accomplish their financial goals.
3. Economic Justice: A determination of principles used to guide the design and development of our economic institutions to ensure participants are able to earn living wages, have access to protections through contracts and agreements, and engage in the exchange of goods and services.
4. Financial Access Deserts: Areas without traditional financial institutions and services, specifically a U.S. Census tract where there are no financial service providers within a ten-mile radius (Morgan, Pinkovskiy, & Yang, 2016).

REFERENCES

Amagir, A., Groot, W., Maassen van den Brink, H., & Wilschut, A. (2017). A Review of Financial-Literacy Education Programs for Children and Adolescents. *Citizenship, Social and Economics Education*, 17(1), 56–80.

Bandura, A. (1977). Self-Efficacy: Toward a Unifying Theory of Behavioral Change. *Psychological Review*, 84(2), 191–215.

Becker, G. S. (1965). A Theory of the Allocation of Time. *Economic Journal*, 75(299), 493–517.

Birkenmaier, J., & Sherraden, M. (2018). Cultural Competence in Financial Counseling and Coaching. http://www.professionalfincounselingjournal.org/cultural-competence-in-financial-counseling-and-coaching.html.

Boyce, L., & Danes, S. M. (1998). *Evaluation of the NEFE High School Financial Planning Program*. Englewood, CO: National Endowment for Financial Education.

Collier, C., Brice, A. E., & Oades-Sese, G. (2007). The Assessment of Acculturation. In G. B. Esquivel, E.C. Lopez, and S. G. Nahari, (Eds.), *Multicultural Handbook of School Psychology: An Interdisciplinary Perspective* (pp. 353–380). Mahwah, NJ: Lawrence Erlabum Associates.

Cooper, D. (2019). Raising the Federal Minimum Wage to $15 by 2024 Would Lift Pay for Nearly 40 Million Workers. Economic Policy Institute, February 5. https://www.epi.org/publication/raising-the-federal-minimum-wage-to-15-by-2024-would-lift-pay-for-nearly-40-million-workers/.

Cordero, J. M., & Pedraja, F. (2019). The Effect of Financial Education Training on the Financial Literacy of Spanish Students in PISA. *Applied Economics*, 51(16), 1679–1693.

Danes, S. (2005). *Evaluation of the NEFE-High School Financial Planning Program 2003–2004*. St Paul: University of Minnesota, Family Social Science Department.

Farrell, L., Fry, T. R. L., & Risse, L. (2016). The Significance of Financial Self-efficacy in Explaining Women's Personal Finance Behavior. *Journal of Economic Psychology*, 54, 85–99.

Fernandes, D., Lynch, J. G., & Netemeyer, R. G. (2014). Financial Literacy, Financial Education, and Downstream Financial Behaviors. *Management Science*, 60(8), 1861–1883.

Financial Consumer Agency of Canada. (2018). National Research Plan for Financial Literacy 2016–2018. https://www.canada.ca/en/financial-consumer-agency/programs/research/national-research-plan-2016-2018.html.

Gale, W. G., Harris, B. H., & Levine, R. (2012). Raising Household Saving: Does Financial Education Work? *Social Security Bulletin*, 72(2), 39–48.

Gale, W. G., & Levine, R. (2010). Financial Literacy: What Works? How Could It Be More Effective? https://www.brookings.edu/wp-content/uploads/2016/06/10_financial_literacy_gale_levine.pdf.

Gartner, K., & Todd, R. M. (2005). Effectiveness of Online Early Intervention Financial Education Programs for Credit-Card Holders. In Federal Reserve Bank of Chicago Proceedings (No. 962).

Halloran, J. (2018). 5 Ways Women Can Better Their Financial Future. *Forbes*, June 11. https://www.forbes.com/sites/forbes-summit-talks/2018/07/11/5-ways-women-can-better-plan-their-financial-future/#5ad9af813b95.

Hastings, J. S., Madrian, B. C., & Skimmyhorn, W. L. (2013). Financial Literacy, Financial Education, and Economic Outcomes. *Annual Review of Economics*, 5, 347–373.

Hathaway, I., & Khatiwada, S. (2008). Do Financial Education Programs Work? Federal Reserve Bank of Cleveland Working Paper, No 08-03. https://www.clevelandfed.org/en/newsroom-and-events/publications/working-papers/working-papers-archives/2008-working-papers/wp-0803-do-financial-education-programs-work.aspx.

Heinrich, C. J. (2013). Targeting Workforce Development Programs: *Who* Should Receive *What* Services? And *How Much*? University of Maryland, School of Public Policy: Center for Policy Exchanges, Atlantic Council. http://www.umdcipe.org/conferences/WorkforceDevelopment/Papers/Workforce%20Development_Heinrich_Targeting%20Workforce%20Development%20Programs.pdf.

Higher Rock Education. (2018). Financial Equity. https://www.higherrockeducation.org/glossary-of-terms/financial-equity.

Institute for Women's Policy Research. (2021). Pay Equity & Discrimination. https://iwpr.org/equal-pay-about/.

Jump$tart Coalition for Personal Financial Literacy. (2006). Financial Literacy Shows Slight Improvement among Nation's High School Students. Rep., Washington, DC.

Kim, K. T., Anderson, S. G., & Seay, M. C. (2018). Financial Knowledge and Short-Term and Long-Term Financial Behaviors of Millennials in the United States. *Journal of Family and Economic Issues*, 40, 194–208. https://doi.org/10.1007/s10834-018-9595-2.

Loke, V., Birkenmaier, J., & Hageman, S. A. (2017). Financial Capability and Asset Building in the Curricula: Student Perceptions. *Journal of Social Work Education*, 53(1), 84–98. https://doi.org/10.1080/10437797.2016.1212751.

Loke, V., Choi, L., & Libby, M. (2015). Increasing Youth Financial Capability: An Evaluation of the MyPath Savings Initiative. *Journal of Consumer Affairs*, 49(1), 97–126.

Lusardi, A., Michaud, P., & Mitchell, O. S. (2017). Assessing the Impact of Financial Education Programs: A Quantitative Model. http://www.pensionresearchcouncil.org.

Lusardi, A., & Mitchell, O. S. (2014). The Economic Importance of Financial Literacy: Theory and Evidence. *Journal of Economic Literature*, 52(1), 5–44.

Lyons, A. C., Palmer, L., Jayaratne, K. S. U., & Scherpf, E. (2006). Are We Making the Grade? A National Overview of Financial Education and Program Evaluation. *Journal of Consumer Affairs*, 40(2), 208–235.

Mandell L. (2008). Financial Literacy of High School Students. In *Handbook of Consumer Finance Research* (pp. 163–183). New York: Springer.

Mandell, L., & Klein, L. S. (2009) The Impact of Financial Literacy Education on Subsequent Financial Behavior. *Journal of Financial Counseling and Planning*, 20(1), 15–24.

Merolla, D. M. (2017). Self-efficacy and Academic Achievement: The Role of Neighborhood Cultural Context. *Sociological Perspectives*, 60(2), 378–393.

Mincer, J., & Polachek, S. (1974). Family Investment in Human Capital: Earnings of Women. *Journal of Political Economy*, 82(2, pt. 2), S76–S108.

Minnesota Psychological Association. (2018). Marginalized Populations. http://www.mnpsych.org/index.php?option=com_dailyplanetblog&view=entry&category=division%20news&id=71:marginalized-populations.

Morgan, D. P., Pinkovskiy, M. L., & Yang, B. (2016). Banking Deserts, Branch Closings, and Soft Information. Liberty Street Economics, Federal Reserve Bank of New York, July 12. https://libertystreeteconomics.newyorkfed.org/2016/03/banking-deserts-branch-closings-and-soft-information.html.

Njai, R., Siegel, P., Yin, S., & Liao, Y. (2017). Prevalence of Perceived Food and Housing Security—15 States, 2013. *Morbidity and Mortality Weekly Report*, 66(1), 12–15. http://dx.doi.org/10.15585/mmwr.mm6601a2.

Piachaud, D. (1987). The Distribution of Income and Work. *Oxford Review of Economic Policy*, 3, 41–61.

Project for Public Spaces. (2007). What Is Placemaking? https://www.pps.org/article/what-is-placemaking.

Serido, J., & Shim, S. (2017). Approaching 30: Adult Financial Capability, Stability, and Well-Being. https://static1.squarespace.com/static/597b61a959cc68be42d2ee8c/t/598a844ecd39c31515c51c7f/1502250072075/APLUS_WAVE4.pdf.

Sherraden, M. S., Frey, J. J., & Birkenmaier, J. (2016). *Financial Social Work: Handbook of Consumer Finance Research*. Cham: Springer.

Shildrick, T., & Rucell, J. (2015). Sociological Perspectives on Poverty. Joseph Rowntree Foundation. https://www.jrf.org.uk/report/sociological-perspectives-poverty.

UC ANR (University of California, Agriculture and Natural Resources). (2019). Money Talks. http://moneytalks.ucanr.edu.

UCSD Extension. (2019). Python Programming. https://extension.ucsd.edu/courses-and-programs/python-programming.

University of Georgia. (2019). Housing & Home Environment. http://extension.uga.edu/topic-areas/money-family-home/housing-home-environment.html.

U.S. Department of Agriculture. (2019). Single Family Housing Programs. https://www.rd.usda.gov/programs-services/all-programs/single-family-housing-programs.

Van Dyke, M. E., Komro, K. A., Shah, M. P., Livingston, M. D., & Kramer, M. D. (2018). State-Level Minimum Wage and Heart Disease Death Rates in the United States, 1980–2015: A Novel Application of Marginal Structural Modeling. *Preventative Medicine*, 112, 97–103.

Varcoe, K., Martin, A., Devitto, Z., & Go, C. (2005). Using a Financial Education Curriculum for Teens. *Journal of Financial Counseling and Planning*, 16(1): 63–71.

Wallace, G. (2017). ISU Extension and Outreach CED Helps Iowa Communities with Housing Needs Assessments. Iowa State University: Extension and Outreach, September 14. https://www.extension.iastate.edu/news/isu-extension-and-outreach-ced-helps-iowa-communities-housing-needs-assessments.

Werhan, C. R. (2013). Family and Consumer Sciences Secondary School Programs: National Survey Shows Continued Demand for FCS Teachers. *Journal of Family and Consumer Sciences*, 105(4), 41–45.

Willis, L. E. (2011). The Financial Education Fallacy. *American Economic Review*, 101(3), 429–434.

Conclusion

Timothy J. Shaffer and Nia Imani Fields

We began this book with Ruby Green Smith's powerful phrase of "vigorous reciprocity" as a way to emphasize the thread within Extension's history and mission that has been committed to relational approaches to the fostering and cultivating of communities across the United States (Smith, 1949, p. ix). The possibilities to have such a community-based and community-driven institution committed to cultivating and sustaining democratic life is a real asset. But it must be recognized as such, especially when we acknowledge that democracy has often excluded entire communities—past and present. And institutionally, this relational approach has not always been as positively received as more traditional research has been for its societal benefits in more abstract and general ways. There have always been these threads and seemingly will continue to be. How we weave them together into a stronger rope or rely primarily on one is the type of issue that this book has sought to address, especially when it comes to the different institutions that comprise the Extension system.

Some of the enduring challenges throughout Extension's history, however, are rooted within the systemic inequalities that have shaped aspects of this democratic institution. The Cooperative Extension Service was created by a 1914 law signed by President Woodrow Wilson. As Eric Yellin has noted in his history about racism in the federal workforce, "The racism of white Americans in the early twentieth century should not surprise us, but the practice of racism still requires exploration and explanation by historians" (Yellin, 2013, p. 6). The year prior to the creation of the Cooperative Extension Service, the federal government had enacted policies to racially segregate federal employees. Secretary of Treasury William McAdoo is famous for speaking about the segregation of federal departmental offices:

> There has been an effort in the departments to remove the causes of complaint and irritation where white women have been forced unnecessarily to sit at desks with colored men. Compulsion of this sort creates friction and race prejudice. Elimination of such friction promotes good feeling and friendship. (Wolgemuth, 1959, p. 167)

One of the challenges for thinking about professionals in government, universities, and other social organizations is the tension between expertise and democratic practices, especially when the technical rationality as exemplified by much of the Progressive Era managerialism permitted such explicit racism because, as President Wilson wrote in a letter about his knowledge of the segregation, "the initiative and suggestion of the heads of the departments" led to these actions (Wolgemuth, 1959, p. 163). President Wilson went on to note that these bureaucratic decisions would remove the "friction, or rather than discontent and uneasiness which had prevailed in many of the departments" (Wolgemuth, 1959, p. 163). In more than simply the federal government, the ideas of managerialism and bureaucratic order were often rooted in practices that came from urban reform movements rather than the "homespun, agrarian tradition" (Mowry, 1972, p. 5). The tensions existed between the ways in which individuals and institutions committed to a model of governance that moved away from localized knowledge and experience to one grounded in technical knowledge (Brint, 1994; Stivers, 2000). The shift from "local democracy, local economy" to a nationalized and uniform way of thinking about the multiple elements of early twentieth-century America—"its political, cultural, and economic institutions and practices"—had impacts on all aspects of life (Eisenach, 1994, pp. 18–19).

Additionally, the United States had entered the "machine age," experiencing a dramatic shift with considerable impact on the lives of people of all walks, especially in rural contexts where farming was central (Jordan, 1994). Like now, the early twentieth century was fraught with different views of democracy and, as Beard and Beard (1930, p. 19) put it, "constantly . . . fac[ing] large questions of choice which cannot be solved by the scientific method alone—questions involving intuitive insight, ethical judgment, and valuation, as of old. Science and machinery do not displace all cultural considerations." The formal establishment of Extension during this period highlighted the commitment to scientific approaches to problems while also being rooted in the relational approaches expressed by Ruby Green Smith and others.

During its founding period, Extension was an institution that was grounded in competing commitments that advanced American life through robust scientific experimentation and implementation as well as through relational efforts to cultivate better citizens. From its beginning and during its first decades, the sense was that productivity was only part of Extension's work. As Director of Extension Work C. W. Warburton put it in 1930, "Better crops, better livestock, better food, better clothes, these are among the objects of extension work. But back of it all, the ultimate purpose is to create better homes, better citizens, better communities, better rural living" (Warburton, 1930, p. 293). The sense that, fundamentally, Extension helped cultivate citizens in community with others for improved life is a powerful statement about its work. But as this volume has highlighted, the disparities with respect to *who* counted as an equal citizen and what *community* meant not only to those within but also to those serving them (whether locally or from the federal government) were very real.

The inequalities of Extension's work throughout its history have often been obfuscated, in part, by the desire to have a narrative that frames its identity through a "highly simplified story of land-grant extension work" (Peters, 2017, p. 73). The unsettling stories, as Peters refers to them, are those that can trouble or surprise us. In the context of this book, the "problem has to do with what it leaves out or obscures" (Peters, 2017, p. 74). In many ways, the 1890 institutions, let alone the 1994 land-grant tribal colleges, are typically a footnote to the 1862s—the universities we often think of when referring to land-grant institutions and/or Extension. The sentiment is captured in many histories of Extension. One such example comes from the University of Massachusetts: "To give—to all the people" (University of Massachusetts, 1971, p. 2). But in many situations, this sense of *all* was exclusionary. As has been noted, "From the beginning, extension programs were tailored to the needs of educated and prosperous white farmers. White schools and agents stubbornly refused to share knowledge with black agents and purposely kept them outside the information loop" (Daniel, 2013, p. 158). This practice became a trend:

> Although African American land-grant schools hosted the Negro Extension Service, they shared little of the [Federal Extension Service] largesse. If the Negro Extension Service had been separate but equal, in 1941, there would have been 1,000 black agents instead of 549 in sixteen southern states. Looked at another way, there were 1,303 white farm operators per white extension agent and 2,781 black farm operators per black agent. In 1941, the total Extension Service budget was $14.8 million for whites and $988,000 for African Americans. Black schools and extension workers carved out zones of autonomy but were beholden to white funding and priorities. Black farmers had few resources to subsidize extension agents, so counties with black agents were scarce and agents were underpaid. As U.S. Commission on Civil Rights interviews in 1964 revealed, separate but equal was a myth, a sick joke. Black extension workers had cramped offices, hand-me-down furniture, scarce telephones, and few typewriters or mimeograph machines. Still, extension and home-demonstration agents made an enormous contribution to the lives of black farm families. (Daniel, 2013, p. 159)

Understanding the ways in which Extension has been shaped by racism has an impact today as those echoes and reverberations continue to impact the institutions and the communities they engage. The 1971 annual report from Massachusetts has one reference to nonwhite populations, unsurprisingly included in the report's section entitled "The Non-Affluent Society." The one statement reads as such: "For most staff members, it was their initial experience in teaching black homemakers, and they had as much to learn as the participants did. State staff went to South End somewhat doubtful of their ability to communicate, relate and be accepted; but the home economist, herself a black, bridged the gap beautifully" (University of Massachusetts, 1971, p. 71). For states without 1890 institutions, the absence of so many stories within these annual reports highlights the important work of acknowledging and understand the challenges that have long faced those committed to engaging with diverse individuals and groups. But even in states where 1890s exist, the disparities with *actually* funding Extension work have highlighted the significant work that still needs to happen (Lee & Keys, 2013). In an article published April 26, 2021, titled "A Debt Long Overdue," there is a striking reminder of how legislative (in)action has shaped the work of land-grant institutions and Extension:

> Between the fiscal years of 1957 and 2007, Tennessee budget records show no funds were directed to Tennessee State University, the state's only public HBCU. According to a report by the Tennessee Office of Legislative Budget Analysis, the state owes the university anywhere between $150 million and $544

> million in unpaid land-grant funds.... If the university had received its allotted funding all these years, it could have ... invested in extension programs. (Weissman, 2021)

Acknowledging the systemic lack of funding is a reminder that inequities within the land-grant system and its funding streams are still very much a live question. As the case of Tennessee highlights, there are well-documented shortfalls that would greatly improve the work of these 1890 institutions.

Attempts to educate and train Extension professionals regarding racism within our institutions and our communities have been conducted through the Extension Committee on Organization and Policy at the national level, with one of the most robust examples being the development of "Coming Together for Racial Understanding," a workshop that has engaged the range of land-grant institutions about the important work of understanding, engaging, and addressing racism through group dialogue (Walcott et al., 2020).[1]

So, what is to be done? This book has illuminated aspects of our institutions that reveal the unflattering truths about the intersection of democratic promise and systemic inequity, but it has also shown a way forward that is more inclusive and engaging of diverse populations. The chapters highlight Extension's commitment to advancing democratic aspirations and possibilities. Yet the idea of Extension being rooted in the distribution of cheap and reasonably healthy food alongside the transfer of research-based knowledge, echoing the famous critique of agribusiness in "Hard Tomatoes, Hard Times" (Hightower, 1978), remains dominant. One of the places where we must continue to reimagine and reclaim the roots of Extension's work is through the practice of democracy in daily life. A serious impediment is the diminishment of this more relational and political work from the organization alongside the elevation of science as the preferred way in which we have something to contribute to people's lives. But more than simply trying to retrieve, we must imagine and reimagine the possibilities of this engaged pillar of an institution that has done so much and has the possibility of serving as a locally based, trusted resource for complex challenges bringing together partners willing to engage collaboratively.

If we take seriously the idea of grassroots engagement, Extension professionals must honestly ask themselves what it means to engage in community education, social justice work, and helping to cultivate and encourage cultural competence. This is especially true if you are reading this and not thinking about "social justice," "cultural competence," or "equality" and "equity" as words applicable to you or those around you. All of our decisions have implications, and Extension has a role in working with others to understand and respond to those challenges and opportunities. The industrialization of agriculture, for example, led to broader cultural impacts that cannot be untangled. For example, the number of black farm operators fell from 272,541 in 1959 to 98,000 in 1970 (Hightower, 1972, p. 12). An even longer view highlights the transformation during the last century. The number of Black farmers peaked in 1920. That year, the USDA counted 925,708 Black farmers, which were 14.34 percent of farmers at the time (Reynolds, 2002, p. 24). In the most recent census of agriculture from 2017, the USDA documented that there were 48,697 Black farmers, roughly 1.4 percent of nearly 3.4 million farmers (United States, 2017, p. 80). While many factors shaped federal action, the ways in which our own institutions have played a role in causing some of the issues we attempt to address in this volume cannot be overlooked. We must see how we, or at least our institutions, have contributed to the challenges we seek to improve today. And when possible, acknowledge the impact—positive or negative.

In many ways, this is the moment to look critically at ourselves. With books such as *How to Be an Antiracist* (Kendi, 2019) and others being critically and popularly read, we are in a moment to examine ourselves—individually and collectively—with an eye toward those practices that hinder or exclude. For Extension, we must have a reckoning ourselves if we see the next century of our work being as meaningful and as impactful as possible. How much of our work is within the existing systems, and how much of it helps to create something new, emerging from what is to make possible what ought to be? Even the inclusion in federal programs of foods eaten by different groups beyond dominant groups could have a real impact on the health of entire communities. Acknowledging our diversity improves our practices, but not without the alteration of existing norms and practices.

And finally, while this book has been a long process, the final phases of the project were impacted by the realities of 2020 and 2021. There was a global social movement, a public health pandemic with respect to COVID-19, and a presidential executive order that set forth the policy of the United States "not to promote race or sex stereotyping or scapegoating" and prohibits federal contractors from inculcating such views in their employees in workplace diversity and inclusion trainings (Office of Federal Contract Compliance Programs, 2020). All of these have real implications on our work and communities, especially as we think about community health, economic challenges and opportunities, and the reality that some of the book's authors have been censored in their federally funded equity work. At the time of this writing, the greatest catastrophe in American history continues to ripple through families, communities, states, and the nation.

As this volume goes to print, nearly one million people are dead from COVID-19 in the United States. The disease continues to spread even with the wide availability (but not wide embrace) of vaccines. Additionally, higher education, especially community engagement, has been dramatically impacted as it relates to how people interacted (Paterson, 2020). Colloquially, we've all been "Zoomed out," and there have been transformations of programs from only face-to-face to online—either temporarily or more permanently.

With all these realities, what will our vital relational work with communities look like on the other side of this harrowing chapter of American history? We don't know exactly, but research suggests that we can more plainly see the contrast between individualistic and communal approaches when dealing with common problems impacting the "common good" (Latemore, 2021; Neblo & Wallace, 2021). In an environment increasingly shaped by affective polarization and political sorting (Levendusky & Stecula, 2021), there are significant challenges but also opportunities when we think about interaction at the grassroots level with an eye toward social justice. Engaging across differences through formal organizations (Boyte, 2021) and through informal but trusted environments (Armour, 2021) points to why Cooperative Extension can be such an asset: it is a known resource to communities and one that is shaped by a commitment to education informed by research and practice. For these reasons, we are hopeful. We must be part of what makes that hope a reality and recognize that we have much to give and much to learn.

NOTE

1. For more on this program, see http://srdc.msstate.edu/civildialogue/.

REFERENCES

Armour, S. (2021). Barbershops and Hair Salons Are Enlisted in Covid-19 Vaccine Push. *Wall Street Journal*, July 4. https://www.wsj.com/articles/barbershops-and-hair-salons-are-enlisted-in-covid-19-vaccine-push-11625410802.

Beard, C. A., & Beard, W. (1930). *The American Leviathan: The Republic in the Machine Age*. New York: Macmillan.

Boyte, H. C. (2021). *Beyond the War Metaphor: The Work of Democracy*. Dayton, OH: Kettering Foundation.

Brint, S. (1994). *In an Age of Experts: The Changing Role of Professionals in Politics and Public Life*. Princeton, NJ: Princeton University Press.

Daniel, P. (2013). *Dispossession: Discrimination against African American Farmers in the Age of Civil Rights*. Chapel Hill: University of North Carolina Press.

Eisenach, E. J. (1994). *The Lost Promise of Progressivism*. Lawrence: University Press of Kansas.

Hightower, J. (1972). Hard Tomatoes, Hard Times: Failure of the Land Grant College Complex. *Society*, 10(1), 10–22. doi:10.1007/bf02695245.

———. (1978). *Hard Tomatoes, Hard Times: The Original Hightower Report, Unexpurgated, of the Agribuiness Accountability Project on the Failure of America's Land Grant College Complex and Selected Additional Views of the Problems and Prospects of American Agriculture in the Late Seventies*. Cambridge, MA: Schenkman Publishing Company.

Jordan, J. M. (1994). *Machine-Age Ideology: Social Engineering and American Liberalism, 1911–1939*. Chapel Hill: University of North Carolina Press.

Kendi, I. X. (2019). *How to Be an Antiracist*. New York: One World.

Latemore, G. (2021). COVID and the Common Good. *Philosophy of Management*, 20(3), 257–269. doi:10.1007/s40926-020-00154-w.

Lee, J. M., & Keys, S. W. (2013). *Land-Grant but Unequal: State One-to-One Match Funding for 1890 Land-Grant Universities*. Washington, DC: Association of Public and Land-Grant Universities. https://www.aplu.org/library/land-grant-but-unequal-state-one-to-one-match-funding-for-1890-land-grant-universities/file.

Levendusky, M. S., & Stecula, D. A. (2021). *We Need to Talk: How Cross-Party Dialogue Reduces Affective Polarization*. New York: Cambridge University Press.

Mowry, G. E. (1972). *The Progressive Era, 1900–20: The Reform Persuasion*. American Historical Association Pamphlets 212. Washington, DC: American Historical Association.

Neblo, M. A., & Wallace, J. L. (2021). A Plague on Politics? The COVID Crisis, Expertise, and the Future of Legitimation. *The American Political Science Review*, 115(4), 1524–1529. doi:10.1017/S0003055421000575.

Office of Federal Contract Compliance Programs. (2020). Executive Order 13950—Combating Race and Sex Stereotyping. Washington, DC. https://www.dol.gov/agencies/ofccp/faqs/executive-order-13950.

Paterson, C. S. (2020). SARS 2 COVID-19 Pandemic and Its Effects on Service Learning and Community Engagement. *Journal of Community Engagement and Higher Education*, 12(3), 81–82.

Peters, S. J. (2017). Recovering a Forgotten Lineage of Democratic Engagement: Agricultural and Extension Programs in the United States. In C. Dolgon, T. D. Mitchell, & T. K. Eatman (Eds.), *The Cambridge Handbook of Service Learning and Community Engagement* (pp. 71–80). New York: Cambridge University Press.

Reynolds, B. J. (2002). *Black Farmers in America, 1865–2000: The Pursuit of Independent Farming and the Role of Cooperatives*. Washington, DC: U.S. Department of Agriculture, Rural Business-Cooperative Service.

Smith, R. G. (1949). *The People's Colleges: A History of the New York State Extension Service in Cornell University and the State, 1876–1948*. Ithaca, NY: Cornell University Press.

Stivers, C. (2000). *Bureau Men, Settlement Women: Constructing Public Administration in the Progressive Era*. Lawrence: University Press of Kansas.

United States. (2017). *2017 U. S. Census of Agriculture* (Vol. 1: Geographic Area Series, Part 51). Washington, DC: U.S. Department of Agriculture.

University of Massachusetts. (1971). *Sixty Years of Cooperative Extension Service in Massachusetts: A History*. Amherst: University of Massachusetts.

Walcott, E., Raison, B., Welborn, R., Pirog, R., Emery, M., Stout, M., Hendrix, L., & Ostrom, M. (2020). We (All) Need to Talk about Race: Building Extension's Capacity for Dialogue and Action. *Journal of Extension*, 58(5), 5COM1. https://archives.joe.org/joe/2020october/comm1.php.

Warburton, C. W. (1930). Six Million Farms as a School. *Journal of Adult Education*, 2(3), 289–293.

Weissman, S. (2021). A Debt Long Overdue. *Inside Higher Ed*, April 26. https://www.insidehighered.com/news/2021/04/26/tennessee-state-fights-chronic-underfunding.

Wolgemuth, K. L. (1959). Woodrow Wilson and Federal Segregation. *Journal of Negro History and Theory*, 44(2), 158–173.

Yellin, E. S. (2013). *Racism in the Nation's Service: Government Workers and the Color Line in Woodrow Wilson's America*. Chapel Hill: University of North Carolina Press.

Contributors

Manami Brown began her University of Maryland, Baltimore City Extension career as a 4-H youth development educator in 1998. Brown was appointed as the city Extension director in 2006 where she manages a diverse group of faculty, staff, and volunteers. Under her leadership, the Extension's multifaceted community-focused programming meets the interests and needs of residents in the Baltimore region and the state of Maryland. As a tenured 4-H educator, Brown has received multiple awards. She also serves as a contributing author of the "Be the E: Entrepreneurship" curriculum and the author of the "4-H Teen Corps: Developing Youth & Adult Leaders to Strengthen Communities" curriculum from the National 4-H curriculum library. She has been a leader in integrating the practices of youth development with those of community development through research-based programs in the city of Baltimore and Dakar, Senegal, West Africa, in the areas of entrepreneurship, workforce readiness, environmental science, and service-learning. Brown holds a BSW from Morgan State University and a MEd from Johns Hopkins University.

Matt Calvert is the Positive Youth Development Institute director at the University of Wisconsin–Madison's Division of Extension. In this role, he provides leadership to 4-H and community youth development programs. Calvert has a PhD in educational policy studies from the University of Wisconsin–Madison. His research and Extension work has focused on including youth voice in organizational and community development.

Helen Cheng is an interdisciplinary marine biologist with over ten years of experience in marine, coastal, and estuarine ecology and in fisheries and coastal policy and management. Helen integrates public outreach and engagement into her research and fosters collaborations and partnerships with groups such as research institutions, federal, state, and local agencies, and grassroots organizations. Recently, Helen worked for New York Sea Grant and the Science and Resilience Institute at Jamaica Bay as a coastal resilience Extension specialist based in New York City. Her work in Extension and outreach included communicating and translating relevant science of climate and weather information with urban coastal communities, developing programs to support community engagement and research efforts to enhance resilience, and informing new science research and decision-making. Helen is currently studying social-ecological systems and management relating to marine and coastal ecosystems and fisheries resources for her PhD at Northeastern University.

Nia Imani Fields is the Maryland 4-H Program leader and assistant director of Maryland Extension. Fields has a doctorate in Urban Affairs and Public Policy from Morgan State University and has a long career in youth and community engagement. As the Maryland 4-H Program leader, Fields provides leadership and direction for 4-H youth development programs, faculty, and staff. Her true purpose in life is to expose as many young people as possible to new and exciting experiences—experiences that encourage youth to dream big.

Latoya M. Hicks serves as the assistant director of civil rights, compliance, inclusion programs, and training administration for the University of Maryland, College of Agriculture and Natural Resources (AGNR), Office of Human Resources Management and Compliance Programs. In such capacity, she provides federally assisted programmatic oversight to areas of research, Extension, and employment and is responsible for developing, maintaining, and evaluating college-wide nondiscrimination policies and procedures for the equitable programmatic delivery to the state of Maryland as a land-grant recipient. Prior to joining AGNR in 2018, Latoya was with the U.S. Department of Agriculture (USDA), National Institute of Food and Agriculture (NIFA), where she served in several roles including as NIFA's limited English proficiency (LEP) program manager providing national guidance, training, advisory, and consulting services. Latoya has more than twenty years of extensive auditing, equal employment opportunity and compliance, diversity and inclusion, human resources, and regulatory experience in various federal agencies and the private sector. She has a DBA and MBA; her certifications include advance mediation, 504 coordinator, Title IX coordinator and administrator, Professional in Human Resources (PHR), Society for Human Resource Management Certified Professional (SHRM-CP), and Senior Certified Affirmative Action and Professional (Sr. CAAP).

Davin Holen is a coastal community resilience specialist and assistant professor for the Alaska Sea Grant Marine Advisory Program and collaborative faculty at the Alaska Center for Climate Assessment and Policy in the International Arctic Research Center, both at the University of Alaska Fairbanks. Davin is a lifelong Alaskan from the Sustina Valley who has a BA in history and MA in anthropology from the University of Alaska Anchorage, and a PhD in anthropology from the University of Alaska Fairbanks. Davin facilitates workshops and other activities related to coastal resilience addressing monitoring, mitigation, and adaptation to local stressors from climatic and ocean changes. Davin is an anthropologist who in his first career conducted social-science research in rural communities from the Arctic to southeast Alaska and continues to work closely with these same communities providing data and decision support tools for communities to make informed decisions about adapting to future climate and environmental scenarios. Davin manages the website Adapt Alaska hosted by Alaska Sea Grant.

Jeff Howard is currently an assistant director with University of Maryland Extension and is the unit head for organization and faculty development. He is the former Maryland State 4-H program leader where he oversaw the 4-H youth development program for eleven years. He is a former trustee with the National 4-H Council and National 4-H Congress and is currently a board member with State 4-H International Exchange. He currently serves on the National 4-H Champions Group for LGBTQ+ Youth Inclusion. His scholarship focus is on international youth programming and also advocacy for inclusion and cultural acceptance with a particular focus on LGBTQ youth.

Shannon Klisch is an academic coordinator with the youth, families, and communities programs of the University of California Cooperative Extension in San Luis Obispo and Santa Barbara Counties. Shannon holds a master of public health degree with a focus on community health education. Her primary areas of focus are food access and community and youth engagement.

Michelle Krehbiel is a youth development specialist and associate professor at the University of Nebraska-Lincoln Extension 4-H Youth Development. Her programming and scholarly research focus on creating positive youth development environments in informal education settings, promoting positive physical and mental health habits in youth, working with diverse youth populations, and conducting alternative program evaluation. She received her PhD from Kansas State University in family studies and human services.

Andrew Lazur is a statewide water quality specialist with the University of Maryland Extension focusing on drinking water quality, private wells, and septic system education. He has been involved in the aquatic science field for over thirty-five years having worked in aquaculture, conservation of aquatic species, pond management, and water quality management.

Erin Ling is a senior water quality Extension associate at Virginia Tech in the Biological Engineering Department. She has a BA from Virginia Tech in international development and two master's degrees from Penn State in environmental pollution control and rural sociology. She has experience with research involving adoption of best management practices for improved water quality and community-based social marketing. Currently she coordinates the Virginia Household Water Quality Program with Virginia Cooperative Extension, which offers affordable water testing and education to Virginia's 1.6 million well and spring users and conducts youth water quality education and applied research.

Lindsey Lunsford is an assistant professor of food systems, education, and policy at Tuskegee University. Lunsford's research focuses on "restorying" African American food systems and foodways in pursuit of cultural justice and food sovereignty. She received her PhD in integrative public policy and development from Tuskegee University. As a scholar activist and agriculture advocate, She is blazing the trail for the upcoming generation of rural and urban agriculturalists. Lunsford ranks nationally as one of the "Top 20 Emerging Leaders in Food and Agriculture" in the United States.

Teresa McCoy, DPA, is director of Learning and Organizational Development at the Ohio State University Extension.

Fe Moncloa is the 4-H youth development advisor at the University of California, Cooperative Extension in Santa Clara County. Fe's programmatic and research projects seek to address the inequities in science literacy among marginalized youth. Fe has a PhD in the social context of education and policy studies from the University of California, Santa Cruz.

Keith Nathaniel has been with UC Cooperative Extension since 1994. He began as program coordinator and later moved on to becoming the 4-H youth development advisor and county director for Los Angeles County. Along with his many years of leadership experience and organizational knowledge, his academic expertise is in youth and adolescent development, youth leadership, civic engagement, and social capital. Keith earned a bachelor's degree in sociology from the University of California, Davis; a master's degree in educational leadership from Florida Agricultural and Mechanical University; and a doctorate in educational leadership from the University of California, Los Angeles.

Megan H. Owens is an assistant professor in the Department of Recreation, Park and Tourism Administration at Western Illinois University. She received her PhD from the University of Illinois, Urbana-Champaign specializing in youth development practices. Megan has worked with community youth and family programming for over twenty years including in the role of Extension state specialist overseeing the research, evaluation, and professional development for Maryland 4-H programs statewide. Megan's research focuses on youth social-emotional learning development through out-of-school-time programs with a particular emphasis on the youth–adult relationship.

Daphne Pee is a science communicator with a background in the environmental sciences. She has worked extensively in the field of water quality, focusing on program development and evaluation. She received a master of environmental management from Duke University.

Scott J. Peters is professor in the Department of Global Development at Cornell University. Situated in the emerging interdisciplinary field of "civic studies," Peters centers his work as a scholar and educator on the project of advancing democratic varieties of public engagement in the academic profession. He is the author of multiple books, most recently *In the Struggle: Scholars and the Fight against Industrial Agribusiness in California*, coauthored with Daniel J. O'Connell. He received his PhD from the University of Minnesota.

Norman E. Pruitt is the director of the University of Maryland, College of Agriculture and Natural Resources, Office of Human Resources Management and Compliance Programs. Norman manages all aspects of legal matters, human resources, diversity, and compliance programs including federal land-grant concerns for the college. Prior to returning to the University of Maryland College Park, Norman worked fourteen years with the U.S. Department of Agriculture (USDA), National Institute of Food and Agriculture (NIFA) where he served as the national program compliance review leader and finished his NIFA career as interim civil rights director. He has conducted reviews in forty states/territories of the United States involving fifty-eight land-grant universities and given over thirty-five invited talks. He established review techniques and designed training to identify the level of participation of underserved, underrepresented, and socially disadvantaged clientele in USDA programs that received federal financial assistance. Prior to joining NIFA Norman was employed for twenty-seven years by the University of Maryland College Park, College of Agriculture and Natural Resources, where he held a faculty Extension appointment and many administrative positions. Norman earned an MBA in finance, MS in agriculture, and BS in journalism from the University of Maryland College Park.

Derrick Robinson is a research economist working in the areas of marine aquaculture, natural resources, labor and worker justice, household resource management, and public policy development. Previously, he has worked at the University of California as a program coordinator for a financial literacy and education program, Money Talks. He serves across multiple levels of influence in partnership with community organizations, stakeholders, and leaders to improve equitable economic benefits for communities through education and advisement around community economic development. Derrick received his PhD in applied economics from Auburn University.

Timothy J. Shaffer is the Stavros Niarchos Foundation Chair of Civil Discourse and associate professor in the Joseph R. Biden, Jr. School of Public Policy and Administration at the University of Delaware. He is also director of civic engagement and deliberative democracy with the National Institute for Civil Discourse at the University of Arizona. In addition to this volume, Shaffer is author or coeditor of five books including *Deliberative Pedagogy: Teaching and Learning for Democratic Engagement* and *Creating Space for Democracy: A Primer on Dialogue and Deliberation in Higher Education.*

Maurice D. Smith Jr. serves as a national program leader within the National Institute of Food and Agriculture (NIFA) Institute of Youth, Family, and Community. In collaboration with the land-grant university system, Cooperative Extension System, federal agencies, and nongovernmental organizations, Maurice provides national leadership for 1890 and 4-H positive youth development programs with an emphasis on minority-serving institutions and underserved youth. Prior to NIFA, Maurice served as assistant professor and Extension specialist with Virginia State University focusing on civic engagement and leadership development. Maurice also served as a 4-H Extension agent with Virginia Cooperative Extension. He received his PhD in agricultural and Extension education from Pennsylvania State University. He received a MS degree from Virginia Tech in agriculture and life sciences and his BS in agriculture with a concentration in agribusiness and economics from Virginia State University. He is a member of the national society of Minorities in Agriculture, Natural Resources and Related Sciences.

Katherine E. Soule is a health equity Cooperative Extension advisor for the University of California. Katherine works across multiple levels of influence in partnership with her community, student leaders, volunteers, and colleagues to improve equity, access, and belonging. She received her PhD from the University of Georgia.

Amanda Wahle is a 4-H youth development specialist for University of Maryland, Extension. Amanda's program work and research focuses on leadership development, environmental science, camping, shooting sports, diversity and inclusion, healthy living, mental health, and community service. She has an MS degree in applied psychology from the University of Baltimore.

Sacoby Wilson is an associate professor with the Maryland Institute for Applied Environmental Health in the University of Maryland, College Park School of Public Health. He directs the Center for Community Engagement, Environmental Justice, and Health (CEEJH). Wilson has over twenty years of experience as an environmental health scientist in the areas of exposure science, environmental justice, environmental health disparities, and community-engaged research, including crowd science and community-based participatory research (CBPR), water quality analysis, air pollution studies, built environment, industrial animal production, climate change, community resiliency, and sustainability. He works primarily in partnership with community-based organizations to study and address environmental justice and health issues and translate research to action. Wilson received his MS degree in 2000 and his PhD in 2005 from University of North Carolina at Chapel Hill. He received his BS from Alabama Agricultural and Mechanical University in 1998.

Robert Zabawa has worked at Tuskegee University for over thirty-six years. He received his PhD in anthropology from Northwestern University. His domestic research and outreach includes small-scale and minority farming systems with an emphasis on land ownership, heir property, family networks, resettlement, and policy focusing on racial/ethnic disparities as it pertains to the U.S. South. His international research includes agricultural development, production decision making, value chain analysis, and marketing strategies. He has worked in Belize, Senegal, and Tanzania, and for the past twenty years he has been on a team promoting the adoption of the orange sweet potato in Ghana. He currently codirects the integrative public policy and development PhD program, and is a Community Resource Development specialist in the Tuskegee University Cooperative Extension Program.